Generalized Estimating Equations

Generalized Estimating Equations

James W. Hardin
Joseph M. Hilbe

CHAPMAN & HALL/CRC

A CRC Press Company
Boca Raton   London   New York   Washington, D.C.

Library of Congress Cataloging-in-Publication Data

Hardin, James W. (James William)
  Generalized estimating equations / James W. Hardin, Joseph M. Hilbe.
    p. cm.
  Includes bibliographical references and index.
  ISBN 1-58488-307-3 (alk. paper)
    1. Generalized estimating equations. I. Hilbe, Joseph. II. Title.

QA278.2 .H378 2002
519.5'36—dc21   2002067404

This book contains information obtained from authentic and highly regarded sources. Reprinted material is quoted with permission, and sources are indicated. A wide variety of references are listed. Reasonable efforts have been made to publish reliable data and information, but the author and the publisher cannot assume responsibility for the validity of all materials or for the consequences of their use.

Neither this book nor any part may be reproduced or transmitted in any form or by any means, electronic or mechanical, including photocopying, microfilming, and recording, or by any information storage or retrieval system, without prior permission in writing from the publisher.

The consent of CRC Press LLC does not extend to copying for general distribution, for promotion, for creating new works, or for resale. Specific permission must be obtained in writing from CRC Press LLC for such copying.

Direct all inquiries to CRC Press LLC, 2000 N.W. Corporate Blvd., Boca Raton, Florida 33431.

Trademark Notice: Product or corporate names may be trademarks or registered trademarks, and are used only for identification and explanation, without intent to infringe.

Visit the CRC Press Web site at www.crcpress.com

© 2003 by Chapman & Hall/CRC

No claim to original U.S. Government works
International Standard Book Number 1-58488-307-3
Library of Congress Card Number 2002067404
Printed in the United States of America 3 4 5 6 7 8 9 0
Printed on acid-free paper

To our wives, Mariaelena Castro-Hardin and Cheryl Lynn Hilbe, and our children, Taylor Antonio Hardin, Conner Diego Hardin, Heather Lynn Hilbe O'Meara, Michael Joseph Hilbe, and Mitchell Jon Hilbe.

Preface

Generalized Estimating Equations is written for the active researcher as well as for the theoretical statistician. Our goal throughout has been to clarify the nature and scope of generalized estimating equations (GEE) and to demonstrate its relationship to alternative panel models.

This text assumes that the reader has a fundamental understanding of generalized linear models (GLM). We shall provide an overview of GLM, but intend it to be merely a review. The more familiar a reader is with GLM, the easier it will be to recognize how the basic GLM algorithm can be extended to incorporate the modeling of longitudinal and clustered data by means of generalized estimating equations.

Generalized Linear Models is essentially a unified method of analyzing certain types of data situations. It is based on the exponential family of probability distributions, which includes the Gaussian or normal, the binomial, Poisson, gamma, inverse Gaussian, geometric, and for a given ancillary parameter, the negative binomial. The binomial models themselves include the logit, probit, log-log, and complementary log-log, among others. Hence, one may use GLM to model OLS regression as well as logistic, probit, and Poisson regression models. The ability to compare parameter estimates, standard errors, and summary statistics between models gives the researcher a powerful means by which he or she may arrive at an optimal model for a given dataset. However, being likelihood based, GLMs assume that individual rows in the data are independent from one another. However, in the case of longitudinal and clustered data, this assumption may fail. The data are correlated. The clustering units are many times called panels; hence their reference as panel data.

Although statisticians created methods within the GLM framework to help correct for correlated data, it became evident that these methods were not sufficient. GEE was explicitly developed to serve as a means to extend the GLM algorithm to accommodate the modeling of correlated data that would have otherwise been modeled using straightforward GLM methods. We note as well that GEE has itself been extended, and at times in a manner that substantially varies from the original GLM approach.

Our intent in writing this text is to provide an overview of the GEE methodology in all of its variations as well as to compare it with other methods that are used to model correlated and clustered data. However, we concentrate our discussion to the general GEE approach.

We have organized the text into four divisions, represented by four main chapters; a fifth chapter lists data and useful programs. The first chapter provides an introduction to the subject matter. The second chapter serves as a review of generalized linear models. We first offer an historical perspective to the development of GLM methodology and point out methods by which the GLM algorithm has been extended to meet particular modeling purposes. We then review basic modeling strategies wherein we focus on the nature and scope of the estimating equation. By focusing attention on the estimating equation of familiar models, we believe it is easier to understand the more complex generalized estimating equation. Finally, we use Chapter 2 to introduce panel data and discuss many of the available likelihood-based models that have been used to deal with such data situations.

Chapter 3 concentrates on the varieties of generalized estimating equations. In fact, we have specifically organized the chapter to facilitate a comparison of the different types of GEE models. The prime division is between marginal or population averaging models and subject specific models. Wherever possible we attempt to demonstrate the source of observed differences in output between different software applications when they occur. Typically they differ because of alternative formulae in the estimating algorithms. Computational variations are usually minor, and involve an extra term in the denominator of an ancillary equation.

Chapter 4 deals with residual analysis and model goodness of fit. We demonstrate many graphical and statistical techniques that can be applied to GEE analysis. Numerous journal articles have recently been published dealing with GEE fit analysis; we attempt to summarize and demonstrate the methods that seem most appropriate. We do recognize, however, that there are as yet few commercial software applications implementing these methods.

We have tried to remain faithful to the title of our text. Notably, we focus our attention to the varieties of GEE models without overly expanding the discussion to include alternative approaches to the modeling of panel data, e.g., hierarchical models, mixed models, and random-effects models. However, we do discuss and show output from some of these alternatives when they are either equivalent or nearly so to the GEE models of primary interest. Ignoring the likelihood-based and simulation-based models would have been shortsighted since we desire the reader to be aware of these available alternative choices.

We perhaps present more mathematical and algorithmic detail than other texts in the area. It is our belief that this approach will be of value to a wider audience. Our goal is to address the needs of the practicing researcher rather than limiting the presentation to the theoretical statistician. However, we hope that the text will be of use to the latter as well. We focus on origins, applications, relationships, and interpretation—all of which we perceive to be useful to the researcher. We try not to present too many theoretical derivations, and we make our presentation in summation notation rather than in matrix notation wherever possible. When matrix results or arguments are required, we include the sizes of matrices to more clearly illustrate the results. Consequently, there is often more explanation than is necessary for the more statistically erudite reader, but we hope that it makes for more meaningful reading and application for those analysts who are not as grounded in statistical theory.

We have gathered a great deal of information related to GEE methodology. To distinguish each approach, we have developed a taxonomy of models. Various labels can be found in the literature, particularly with respect to GEE extensions. We attempt to adopt those published labels where reasonable. However, because of the variation found in the literature, we have created a common taxonomy related to all relevant models. Care should be taken when reading original articles to understand labels by means of context. As in all aspects of our life, care and common sense should dictate.

In attempting to illustrate as many techniques as possible, we occasionally include examples of fitting models that are not the best choice for the data in use. We fit these "wrong" models for the pedagogical purpose of illustrating techniques and algorithms even though these examples sacrifice correct modeling strategies. We hope the readers will forgive these transgressions on our part.

We wish to recognize many who have contributed to the ideas expressed in this text. John Nelder has been our foremost influence. Others who we consider most important to our efforts include Scott Zeger, Kung-Yee Liang, Roger Newson, Raymond J. Carroll, H. Joseph Newton, Vince Carey, Henrik Schmiediche, Norman Breslow, Berwin Turlach, Gordon Johnston, Thomas Lumley, Bill Sribney, the Department of Statistics faculty at Texas A&M University, and a host of others. We also wish to thank Helena Redshaw, supervisor of the editorial project development of Chapman & Hall/CRC Press, for her encouragement and support for this project.

At Chapman & Hall/CRC Press, we thank Marsha Hecht, Michele Berman, and Jasmin Naim for providing editorial guidance, arranging reviews, and keeping us on schedule. Finally, we express our gratitude and appreciation to Kirsty Stroud, Chapman & Hall/CRC statistics editor, for her initiation, confidence, and support throughout this project.

J.W.H.
J.M.H.

Datasets from this book are available in tab-delimited plain text format from:
http://www.crcpress.com/e_products/downloads/download.asp?cat_no=C3073

Contents

1	**Introduction**	**1**
	1.1 Notational conventions	2
	1.2 A short review of generalized linear models	3
	1.2.1 Historical review	3
	1.2.2 Basics	6
	1.2.3 Link and variance functions	8
	1.2.4 Algorithms	9
	1.3 Software	11
	1.3.1 S-PLUS	12
	1.3.2 SAS	13
	1.3.3 Stata	13
	1.3.4 SUDAAN	14
	1.4 Exercises	15
2	**Model Construction and Estimating Equations**	**17**
	2.1 Independent data	17
	2.1.1 The FIML estimating equation for linear regression	18
	2.1.2 The FIML estimating equation for Poisson regression	21
	2.1.3 The FIML estimating equation for Bernoulli regression	22
	2.1.4 The LIML estimating equation for GLMs	24
	2.1.5 The LIMQL estimating equation for GLMs	27
	2.2 Estimating the variance of the estimates	28
	2.3 Panel data	32
	2.3.1 Pooled estimators	33
	2.3.2 Fixed-effects and random-effects models	34
	2.3.2.1 Unconditional fixed-effects models	35
	2.3.2.2 Conditional fixed-effects models	36
	2.3.2.3 Random-effects models	42
	2.3.3 Population-averaged and subject-specific models	49
	2.4 Estimation	50
	2.5 Summary	50
	2.6 Exercises	52
3	**Generalized Estimating Equations**	**55**
	3.1 Population-averaged (PA) and subject-specific (SS) models	55
	3.2 The PA-GEE for GLMs	57

	3.2.1	Parameterizing the working correlation matrix	58
		3.2.1.1 Exchangeable correlation	59
		3.2.1.2 Autoregressive correlation	66
		3.2.1.3 Stationary correlation	68
		3.2.1.4 Nonstationary correlation	71
		3.2.1.5 Unstructured correlation	72
		3.2.1.6 Fixed correlation	73
		3.2.1.7 Free specification	73
	3.2.2	Estimating the scale variance (dispersion parameter)	76
		3.2.2.1 Independence models	77
		3.2.2.2 Exchangeable models	82
	3.2.3	Estimating the PA-GEE model	85
	3.2.4	Convergence of the estimation routine	89
	3.2.5	ALR: Estimating correlations for binomial models	89
	3.2.6	Summary	93
3.3	The SS-GEE for GLMs		95
	3.3.1	Single random-effects	96
	3.3.2	Multiple random-effects	98
	3.3.3	Applications of the SS-GEE	99
	3.3.4	Estimating the SS-GEE model	103
	3.3.5	Summary	104
3.4	The GEE2 for GLMs		104
3.5	GEEs for extensions of GLMs		106
	3.5.1	Generalized logistic regression	106
	3.5.2	Cumulative logistic regression	108
3.6	Further developments and applications		110
	3.6.1	The PA-GEE for GLMs with measurement error	110
	3.6.2	The PA-EGEE for GLMs	117
	3.6.3	The PA-REGEE for GLMs	119
3.7	Missing data		122
3.8	Choosing an appropriate model		128
3.9	Summary		131
3.10	Exercises		134

4 Residuals, Diagnostics, and Testing — 137

4.1	Criterion measures		139
	4.1.1	Choosing the best correlation structure	139
	4.1.2	Choosing the best subset of covariates	142
4.2	Analysis of residuals		142
	4.2.1	A nonparametric test of the randomness of residuals	143
	4.2.2	Graphical assessment	143
	4.2.3	Quasivariance functions for PA-GEE models	154
4.3	Deletion diagnostics		158
	4.3.1	Influence measures	159
	4.3.2	Leverage measures	165
4.4	Goodness of fit (population-averaged models)		165

		4.4.1 Proportional reduction in variation	165
		4.4.2 Concordance correlation	166
		4.4.3 A χ^2 goodness of fit test for PA-GEE binomial models	167
	4.5	Testing coefficients in the PA-GEE model	169
		4.5.1 Likelihood ratio tests	170
		4.5.2 Wald tests	172
		4.5.3 Score tests	174
	4.6	Assessing the MCAR assumption of PA-GEE models	174
	4.7	Summary	177
	4.8	Exercises	179
5	**Programs and Datasets**		**181**
	5.1	Programs	181
		5.1.1 Fitting PA-GEE models in Stata	182
		5.1.2 Fitting PA-GEE models in SAS	183
		5.1.3 Fitting PA-GEE models in S-PLUS	184
		5.1.4 Fitting ALR models in SAS	185
		5.1.5 Fitting PA-GEE models in SUDAAN	186
		5.1.6 Calculating QIC in Stata	187
		5.1.7 Calculating QICu in Stata	188
		5.1.8 Graphing the residual runs test in S-PLUS	189
		5.1.9 Using the fixed correlation structure in Stata	190
		5.1.10 Fitting quasivariance PA-GEE models in S-PLUS	191
	5.2	Datasets	192
		5.2.1 Wheeze data	192
		5.2.2 Ship accident data	194
		5.2.3 Progabide data	196
		5.2.4 Simulated logistic data	202
		5.2.5 Simulated user-specified correlated data	209
		5.2.6 Simulated measurement error data for the PA-GEE	212

References   **215**

Author index   **219**

Subject index   **221**

CHAPTER 1

Introduction

In this text we address the general field of panel data analysis including longitudinal data analysis, but our main focus is on those models generally classified as generalized estimating equations, or GEE. Throughout, we have endeavored to remain consistent in our use of terms and notation defined in the following paragraphs. Employing strict definitions to these terms will enable the reader to traverse the relevant subject literature.

All GEE models consider an estimating equation that is written in two parts. The first part estimates the regression parameters, and the second estimates the association parameters or the parameters of the second order variance distribution. We present below a schema of the various categories of GEE models. The remainder of the text is devoted to filling in the details.

GEE1   Any GEE model that assumes orthogonality of the estimating equations for the regression and association parameters.

PA   Population-averaged model focusing on the marginal distribution of the outcome.

PA-GEE   A GEE model using moment estimates of the association parameters based on Pearson residuals.

ALR   A GEE model using logistic regression of the odds ratios to estimate the association parameters.

PA-EGEE   A GEE model using the extended quasilikelihood as its genesis rather than the quasilikelihood. The model can use either Pearson residuals or odds ratios for the association parameters.

PA-REGEE   A resistant GEE model using downweighting to remove influence of outliers from the estimation. The model can use either Pearson residuals or odds ratios for the association parameters.

SS   Subject-specific model.

SS-GEE   A GEE model assuming a parametric distribution for the random component and modeling the entire population distribution rather than the marginal distribution.

GEE2   Any GEE model that does not assume orthogonality of the estimating equations for the regression and association parameters.

1.1 Notational conventions

Throughout the text, we use the following acronyms according to the given descriptions:

FIML	Full information maximum likelihood
LIML	Limited information maximum likelihood
LIMQL	Limited information maximum quasilikelihood
RE	Random effects
FE	Fixed effects
GEE	Generalized estimating equation
GEE1	GEE application where the estimating equation for the second-level variance parameters is ancillary and assumed to be orthogonal to the estimating equation of the regression coefficients
GEE2	GEE application where estimating equation for the second-level variance parameters is not assumed to be orthogonal to the estimating equation of the regression coefficients
PA-GEE	GEE-constructed model focusing on the marginal distribution (also known as a *population-averaged* or marginal model)
SS-GEE	GEE-constructed model focusing on the individuals (also known as a *subject-specific* model)
PA-EGEE	GEE-constructed binomial model focusing on the marginal distribution (also known as a *population-averaged* or marginal model) that provides simultaneous estimation of coefficients and association parameters. This technique differs from PA-GEE in the manner in which the association parameters are estimated
PA-REGEE	Resistant GEE-constructed model focusing on the marginal distribution (also known as a *population-averaged* or marginal model) where the model downweights the data to remove influence

We also use the following notation:

$L()$	Likelihood function.
$\mathcal{L}()$	Log-likelihood function.
$\mathcal{Q}()$	Quasilikelihood function.
$\mathcal{Q}^+()$	Extended quasilikelihood function.
$\Psi()$	The estimating equation or generalized estimating equation.

1.2 A short review of generalized linear models

Generalized Estimating Equations (GEE), the prime subject of this text, is traditionally presented as an extension to the standard array of Generalized Linear Models (GLMs) as initially constructed by Wedderburn and Nelder in the mid-1970s. As such, we shall provide an overview of GLM and discuss the various ways that the GLM algorithm has been extended to allow the modeling of correlated data.

1.2.1 Historical review

Peter McCullagh of the University of Chicago and John Nelder of the Imperial College of Science and Technology, London, authored the seminal work on Generalized Linear Models in 1983, in a text with the same name. Major revisions were made in McCullagh and Nelder (1989), which is still the most current edition. This text remains the mainstay and most referenced book on the topic. More importantly, for our purposes, it is the basis upon which Liang and Zeger (1986) introduced a method for handling correlated longitudinal and clustered data.

As likelihood-based models, GLMs are based on the assumption that individual subjects or observations are independent. This assumption is commonly referred to as the *iid* requirement; i.e., observations are independent and identically distributed. However, there are many common data situations for which responses are correlated. For instance, consider a dataset consisting of patient records taken from various hospitals within a state or province. Also, suppose that the data of interest relate to a certain type of medical procedure. It is likely that each hospital has its own treatment protocol such that there is a correlation of treatment effects within hospitals that is absent between hospitals. When such a condition exists, the individual data records are not independent, and, hence, violate the iid assumption upon which many likelihood and quasilikelihood models are based.

In the late 1970s, John Nelder designed the first commercial software developed exclusively for GLMs. Called GLIM, for Generalized Linear Interactive Modeling, the software was manufactured and distributed by Numerical Algorithms Group in Great Britain.

Later, Nelder and GLIM team members introduced capabilities into GLIM that allowed adjustment of the variance-covariance or Hessian matrix so that the effects of extra correlation in the data would be taken into account with respect to standard errors. This was accomplished through estimation of the dispersion statistic. There are two types of dispersion statistics in GLM methodology. The first type is based on the deviance statistic; the second on the Pearson χ^2 statistic. As we discuss later, the overall model deviance and Pearson χ^2 statistics are summary measures of model fit that are traditionally included in model output. The deviance dispersion is derived by dividing the deviance statistic by the model residual degrees of freedom. Likewise, the Pearson χ^2 statistic is calculated by dividing the summary Pearson χ^2 by the same model degrees of freedom. The residual degrees of freedom is itself defined as $(n-p)$ where n is the number of cases in the model and p refers to the number of model predictors, including a constant if applicable.

Depending on the type of correlation effect, we characterize response data on counts and binomial trials as underdispersed or overdispersed. If we are to more appropriately model such data, we must amend the usual GLM and estimating algorithm to address the correlation effects.

The earliest method used to adjust standard errors due to perceived correlation effects was to divide each parameter standard error by the square root of either the deviance-dispersion or Pearson χ^2 dispersion. This procedure is called the scaling of standard errors. It is a post-estimation adjustment of standard errors that has no effect on the fitted regression coefficients.

For binomial and count models estimated using GLM methodology, a dispersion statistic greater than 1.0 indicates possible extra correlation in the data. Scaling is an attempt to adjust the standard errors to values that would be observed if the data were not overdispersed. That is, scaling provides standard errors that would be obtained if the dispersion statistic were 1.0.

The above description of scaling is somewhat naive, as we shall see. However, the idea behind scaling is to use straightforward model statistics to accommodate data that are marginally correlated. This method still proves useful to the analyst as a first-run look at the data.

We should mention at this point that there are occasions when a model may appear to be overdispersed when in fact it is not. For instance, if the deviance-based dispersion of a Poisson model is greater than 1.0, this provides *prima facie* evidence that the model may be overdispersed. In practice, analysts typically start terming a Poisson model as overdispersed when the dispersion statistic is greater than 1.5 and the number of cases in the model is large. Just how much greater than 1.5 and just how large of a dataset depend on the number of predictors, predictor profile, and the pattern of covariates in the model. Hence, there is no definitive dispersion value over which a model is specifically categorized as overdispersed.

In addition to the above caveat regarding model overdispersion, a model that otherwise appears to be overdispersed may in reality be what we call apparently overdispersed. Apparent overdispersion results when a model omits relevant explanatory predictors, or when the data contain influential and possibly mistakenly coded outliers, or when the model has failed to account for needed interaction terms, or when one or more predictors need to be transformed to another scale, or when the assumed linear relationship between the response and predictors is in fact some other relationship. When any of the above situations occurs, the result may be an inflation of the dispersion statistic. Applying remedies to accommodate the above conditions may result in a value of the reestimated dispersion statistic to be reduced to near 1.0. When this occurs, the original model is proven to have been apparently overdispersed.

On the other hand, if one tests for or makes appropriate changes in the model and the dispersion statistic is still high, then it is likely that the dispersion is real. Other checks may be used to assess overdispersion including the comparison of the mean and variance of the response, or evaluation of residuals.

The important point is that signs of model overdispersion must be evaluated; and if overdisperson is found to be real it must be dealt with in an appropriate manner. The manner in which overdispersion is dealt with in large part depends on the perceived source of the overdispersion, which itself represents excess correlation in the data. Standard methods include scaling, using robust variance estimators, or implementing models that internally adjust for correlated data.

Scaling standard errors is a *post hoc* method of analyzing correlated data. It is performed after the model has been estimated, and only adjusts standard errors. It has no effect on parameter estimates. As such, the major deficiency is that it does not capture, or appropriately adjust for, an identified cluster or correlation effect. The method simply provides an overall adjustment.

Another method that applies an overall adjustment to standard errors has also found favor in a number of scientific disciplines. This method, an alternative variance estimate, has been called by various names over the past several decades, many times depending on the academic discipline employing it. We shall simply refer to it as the sandwich variance estimator. Over time, other related variance estimators have been proposed to more directly address nonindependence, and we discuss one general modification in particular. These alternative variance estimators represent a more sophisticated approach to adjusting inference than simply scaling the standard errors based on the dispersion statistic. However, the adjustment is still *post hoc* and only affects the standard errors, not the parameter estimates themselves.

In the mid-1980s, researchers at Johns Hopkins Hospital in Baltimore developed methods to deal with longitudinal and cluster data using the GLM format. In so doing, they created a 2-step algorithm that first estimates a straightforward GLM, and then calculates a matrix of scaling values. The scaling matrix adjusts the Hessian matrix at the next algorithm iteration. Each subsequent iteration in the algorithm updates the parameter estimates, the adjusted Hessian matrix, and a matrix of scales. Liang and Zeger (1986) provided further exposition of how the matrix of scales could be parameterized to allow user control over the structure of the dependence in the data.

Although this is the barest description of their method, hopefully it illustrates the logic behind the initial versions of the extended GLMs introduced through generalized estimating equations. The method arose to better address the dependence of longitudinal and clustered data. As should be expected, the original GEE algorithm served as a springboard for the development of other methods for dealing with correlated data.

Because GEE is traditionally presented as an extension of generalized linear models, we outline the various features that characterize a GLM. A much more thorough examination can be found in Hardin and Hilbe (2001).

1.2.2 Basics

Many of the models that have now been incorporated under the rubric Generalized Linear Models (GLMs) were previously (and many still are) estimated using maximum likelihood methods. Examples include logistic regression, Poisson regression, and probit regression. Each of these regression routines were in use prior to the creation of the GLM algorithm. Why duplicate what was already available?

In the early 1970s, computing was usually performed on mainframe computers. Academics could purchase execution time on campus computers, typically located within the newly developing Departments of Computer Science. Sometimes researchers and analysts were fortunate to have easy access to computing facilities; but that was rather rare. Computer use was absolutely necessary in order to estimate parameters using maximum likelihood optimization techniques. The simple matrix inversion of the Hessian required for maximum likelihood algorithms is not simple at all if one has to calculate the inverse by hand. Moreover, maximum likelihood optimization algorithms require tractable starting values and substantial computing power, especially for large datasets.

There was a clear need to find an optimization method by which otherwise nonlinear models could be estimated using standard OLS methods. Wedderburn and Nelder discovered that the methods used to estimate weighted linear regression could be adjusted to model many data situations that were previously estimated via maximum likelihood, particularly for those maximum likelihood models based on the exponential family of distributions. They accomplished this by applying the Iterative Weighted Least Squares (IWLS) algorithm already in use. In addition, they employed a link function which linearized such functions as the logistic, probit, and log. The IWLS algorithm was later renamed IRLS, meaning Iterative Re-weighted Least Squares to emphasize the updating step for the weights in the algorithm. Also, it was renamed to distinguish it from the traditional weighted least squares WLS algorithm. Hardin and Hilbe (2001) point out that the name change is not without some etymological controversy; Nelder felt that "reweighted" put too much emphasis on the updating of the weights in the OLS calculation given that the synthetic dependent variable is also updated.

Despite some reservations to the name change of the algorithm, IRLS became a common framework for estimating models derived from the exponential family of probability distributions. The algorithm takes advantage of the form of the variance estimate available from Fisher scoring to develop an easy framework from which computer code can be developed. Later, when computing memory and processor speed became more available, GLM algorithms were extended to incorporate varieties of Newton–Raphson based estimation. This allowed more complex models to be estimated within an expanded GLM framework.

Generalized linear models, as previously mentioned, are based on the exponential family of distributions. Members of this family include the Gaussian or normal, binomial, gamma, inverse Gaussian, Poisson, geometric, and the negative binomial for a specified ancillary parameter. Liang and Zeger's GEE extension of GLM focused on the traditional Gaussian, binomial, gamma, and Poisson family members, though their application clearly extends to other members.

All members of the traditional class of generalized linear models are based on one of the above probability functions. The likelihood function is simply a re-parameterization of the probability function or density. A probability function estimates a probability based on given location and scale parameters. A likelihood function, on the other hand, estimates the parameters on the basis of given probabilities or means. The idea is that the likelihood estimates parameters that make the observed data most probable or likely. Statisticians use the log transform of the likelihood, however, because it is (usually) more tractable to use in computer estimation. More detailed justification can be found in Gould and Sribney (1999).

Members of the exponential family of distributions have the unique property that their likelihood formulation may be expressed as

$$\exp\left\{\frac{y\theta - b(\theta)}{a(\phi)} - c(y, \phi)\right\} \quad (1.1)$$

For instance, consider the Poisson probability function

$$f(y; \mu) = \frac{e^{-\mu}\mu^y}{y!} \quad (1.2)$$

We may rewrite this function in exponential family form as

$$f(y; \mu) = \exp\left\{\frac{y\ln(\mu) - \mu}{1} - \ln\Gamma(y+1)\right\} \quad (1.3)$$

As mentioned previously there are a number of distributions for which the associated likelihood follows this general form. The power of GLM lies in the ability to develop or derive techniques, statistics, and properties for the entire group simply based on the form of the likelihood.

The expected value of the exponential family distribution is related to the outcome variable of interest. There is a natural connection between these two quantities that allows us to introduce covariates into the model in place of the expected value. This connection is the θ parameter. When a particular distribution is written in exponential family form, the θ parameter is represented by some monotonic differentiable function of the expected value μ. This function links the outcome variable y to the expected value μ. The particular function that results from writing a distribution in exponential form is called the canonical link. In general, we can introduce covariates into the model through *any* monotonic differentiable link function, though we can encounter numeric difficulties if the function fails to enforce range restrictions that define the particular distribution of the exponential family.

For any member distribution of the exponential family of distributions, there is a general link function, called the canonical link, that relates the linear predictor $\eta = \mathbf{X}\beta$ to the expected value μ. These canonical links occur when $\theta = \eta$. For the Poisson model, we see that $\theta = \ln(\mu)$, implying that the canonical link is given by the log-link $\eta = \ln(\mu)$. Since there is no compelling reason that the systematic components of the model should be linear on the scale of the canonical link, we can, as previously mentioned, choose any monotonic differentiable function.

Subsequent to introducing this class of regression models, Wedderburn (1974) showed that the theoretical results could be justified through an assumption of independence of the observations and an assumption that the variance could be written as a function of the mean (up to a scale factor). This set of assumptions is much less conservative than the original assumption of particular parametric distributions. As a consequence, the class of GLMs allows not only a specification of the link function relating the outcome to the covariates, but also a specification of the form of the variance in terms of the mean. These two choices are not limited to specific distributions in the exponential family. Substituting a given link function and variance function into the IRLS algorithm implies a quasilikelihood. If the link and variance functions coincide with choices for a particular distribution of the exponential family, the quasilikelihood is a likelihood proper.

1.2.3 Link and variance functions

There are a number of standard choices in the data analyst's toolbox for specifying the relationship of the expected value of the outcome variable to the linear combination of covariates $\mathbf{X}\beta$. Usually, these choices are driven by the range and nature of the outcome variable. For instance, when the outcome is binary, analysts naturally choose inverse link functions that map any possible calculation of the linear combination of the covariates and associated parameters to a range $(0,1)$ implied by the outcome. The inverse link function is what converts the linear predictor $\mathbf{X}\widehat{\beta}$ into an estimate of the expected value μ. Positive outcomes similarly lead analysts to choose inverse link functions that transform the linear predictor $\eta = \mathbf{X}\beta$ to positive values.

Some standard choices of link and inverse link functions are listed in Table 1.1. Variance functions corresponding to member distributions in the exponential family are listed in Table 1.2.

Other common choices for link functions include the general power link function (which includes the log, reciprocal, and inverse square as special cases) and the odds power link. See Hardin and Hilbe (2001) for a more complete list of link functions and variance functions along with useful expressions for derivatives and range restrictions.

Confusion can arise in reading various texts on GLMs. The link function is the function that converts the expected value μ (which may be range restricted) to the unrestricted linear predictor $\mathbf{X}\beta$. The function is invertible, and often texts will list the inverse link function instead of, or as well as, the link function. Terminology to differentiate these functions derives from the associated name of the link function. For example, with a positive outcome

Link Name	Link function $\eta = g(\mu)$	Inverse Link $\mu = g^{-1}(\eta)$
Complementary log-log	$\ln\{-\ln(1-\mu)\}$	$1 - \exp\{-\exp(\eta)\}$
Identity	μ	η
Inverse Square	$1/\mu^2$	$1/\sqrt{\eta}$
Log	$\ln(\mu)$	$\exp(\eta)$
Log-log	$-\ln\{-\ln(\mu)\}$	$\exp\{-\exp(-\eta)\}$
Logit	$\ln\left(\mu/(1-\mu)\right)$	$e^\eta/(1+e^\eta)$
Probit	$\Phi^{-1}(\mu)$	$\Phi(\eta)$
Reciprocal	$1/\mu$	$1/\eta$

Table 1.1 *Standard link and inverse link functions.*

Distribution	Variance $V(\mu)$
Bernoulli	$\mu(1-\mu)$
Binomial(k)	$\mu(1-\mu/k)$
Gamma	μ^2
Gaussian	1
Inverse Gaussian	μ^3
Negative binomial	$\mu + k\mu^2$
Poisson	μ

Table 1.2 *Variance functions for distributions in the exponential family.*

variable, textbook authors may choose the log-link such that $\ln(\mu) = \mathbf{X}\beta$ while still mentioning the inverse link $\boldsymbol{\mu} = \exp(\mathbf{X}\beta)$.

1.2.4 Algorithms

The estimates of the GLM regression parameter vector β are the solution of the estimating equation given by

$$\frac{\partial \mathcal{L}}{\partial \beta} = 0 \tag{1.4}$$

where $\mathcal{L}$ is the log-likelihood for the exponential family. A Taylor series expansion about the solution β^* is given by

$$0 = \frac{\partial \mathcal{L}}{\partial \beta}(\beta^*) - (\beta - \beta^*)\frac{\partial^2 \mathcal{L}}{\partial \beta \partial \beta^{\mathrm{T}}} + \ldots \tag{1.5}$$

such that a recursive relationship for finding the solution is

$$\beta^{(r)} = \beta^{(r-1)} + \left[-\frac{\partial^2 \mathcal{L}}{\partial \beta \partial \beta^T} \left(\beta^{(r-1)} \right) \right]^{-1} \frac{\partial \mathcal{L}}{\partial \beta} \left(\beta^{(r-1)} \right) \qquad (1.6)$$

The method of Fisher scoring substitutes the expected value of the Hessian matrix, resulting in

$$\beta^{(r)} = \beta^{(r-1)} + \left[E \left(\frac{\partial \mathcal{L}}{\partial \beta} \frac{\partial \mathcal{L}}{\partial \beta^T} \right) \left(\beta^{(r-1)} \right) \right]^{-1} \frac{\partial \mathcal{L}}{\partial \beta} \left(\beta^{(r-1)} \right) \qquad (1.7)$$

Filling in the details, the updating steps are defined by

$$\left\{ \sum_{i=1}^n \frac{1}{V(\mu_i) a(\phi)} \left(\frac{\partial \mu}{\partial \eta} \right)_i^2 x_{ji} x_{ki} \right\}_{p \times p} \beta^{(r)}_{p \times 1} =$$

$$\left\{ \sum_{i=1}^n \frac{1}{V(\mu_i) a(\phi)} \left(\frac{\partial \mu}{\partial \eta} \right)_i^2 \left\{ (y_i - \mu_i) \left(\frac{\partial \eta}{\partial \mu} \right)_i + \eta_i \right\} x_{ji} \right\}_{p \times 1} \qquad (1.8)$$

for $j = 1, \ldots, p$. Here, we note that

$$\mathbf{W} = \mathrm{Diag} \left\{ \frac{1}{V(\mu) a(\phi)} \left(\frac{\partial \mu}{\partial \eta} \right)^2 \right\}_{(n \times n)} \qquad (1.9)$$

$$\mathbf{Z} = \left\{ (y - \mu) \left(\frac{\partial \eta}{\partial \mu} \right)_i + \eta \right\}_{(n \times 1)} \qquad (1.10)$$

so that we may rewrite the updating step in matrix notation (with $\mathbf{X}_{n \times p}$) as the weighted OLS equation

$$(\mathbf{X}^T \mathbf{W} \mathbf{X}) \beta^{(r)} = \mathbf{X}^T \mathbf{W} \mathbf{Z} \qquad (1.11)$$

Hence, the IRLS algorithm is succinctly described by the algorithm

1. Initialize the vector of expected values $\boldsymbol{\mu}$.
2. Calculate the linear predictor using the link function $\boldsymbol{\eta} = g(\boldsymbol{\mu})$.
3. Initialize the scalars OldDeviance and NewDeviance to zero.
4. Initialize the scalar tolerance to $1e-6$ (or another small tolerance value), and the scalar DeltaDeviance to one (or some other value larger than tolerance).
5. If |DeltaDeviance| > tolerance then stop.
6. Calculate the weight matrix $\mathbf{W}$.
7. Calculate the synthetic dependent variable $\mathbf{Z}$.
8. Regress $\mathbf{Z}$ on $\mathbf{X}$ using OLS with weights $\mathbf{W}$ to get $\beta^{(r)}$.
9. Calculate the linear predictor $\boldsymbol{\eta} = \mathbf{X} \beta^{(r)}$ from the regression results.
10. Calculate the expected values using the inverse link $\boldsymbol{\mu} = g^{-1}(\boldsymbol{\eta})$.
11. Set OldDeviance equal to NewDeviance.
12. Calculate the deviance and store in NewDeviance.
13. Store into DeltaDeviance the difference of NewDeviance and OldDeviance.
14. Return to step 5.

The substitution given by the method of Fisher scoring admits the use of weighted OLS to iteratively define weights **W** and the synthetic dependent variable **Z**. Later theoretical developments relaxed the assumption that the estimating equation must be defined as the derivative of a log-likelihood.

If we maintain the assumption that the estimating equation is the derivative of the log-likelihood, we can skip the substitution of the expected Hessian matrix given by the method of Fisher scoring. In this approach, we use a Newton–Raphson algorithm to iterate to the maximum likelihood estimates of β. This algorithm is succinctly described by the following algorithm

1. Initialize the coefficient vector β.
2. Calculate the log-likelihood $\mathcal{L}$ for the initial β.
3. Set the scalar `BetaTol` and the scalar `LikTol` to a desired tolerance level.
4. Set the old log-likelihood value $\mathcal{L}_{\text{old}}$ to $\mathcal{L}+2\texttt{LikTol}$ (or some other large value).
5. Initialize the coefficient vector β_{old} to $10\beta+1$ (or some other large values).
6. If $\|\beta - \beta_{\text{old}}\| > \texttt{BetaTol}$ or $|\mathcal{L} - \mathcal{L}_{\text{old}}| > \texttt{LikTol}$ then stop.
7. Calculate the gradient $\mathbf{g} = \partial \mathcal{L}/\partial \beta$ evaluated at β_{old}.
8. Calculate the Hessian $\mathbf{H} = -\partial^2 \mathcal{L}/(\partial \beta \partial \beta^{\mathrm{T}})$ evaluated at β_{old}.
9. Set $\beta_{\text{old}} = \beta$.
10. Set $\mathcal{L}_{\text{old}} = \mathcal{L}$.
11. Calculate the new coefficient vector $\beta = \beta_{\text{old}} + \mathbf{H}^{-1}\mathbf{g}$.
12. Calculate the new log-likelihood $\mathcal{L}$.
13. Return to step 6.

The complete theory and derivation of these algorithms are given in Hardin and Hilbe (2001).

1.3 Software

There are a number of general purpose statistical packages that offer different levels of support for fitting the models described in this text. In the following subsections, we discuss only those packages that were used in preparing the output for the various examples we have used. While we specifically give information and show output from these particular packages, we should emphasize that none of these products fully supports all of the models and diagnostics that we discuss. Each of the packages mentioned in this section has built in support for user-written programs, so that included output and examples can be obtained with sometimes minimal programming effort.

A researcher who intends to investigate all of the details and techniques outlined here will ultimately need to engage in programming. For the diagnostics and tests, the level of programming is minimal so that experience with any of these packages should allow the interested reader sufficient expertise to obtain results. A more profound level of programming expertise is required for models that are extensions from the principal collection of models.

The commercial vendors for the software packages used in this text can provide license information as well as technical support. Depending on the breadth of analyses an individual must perform, having access to more than one of these packages may be a great advantage. Horton and Lipsitz (1999) present a particularly useful review of software that focuses on the four software packages used. Hilbe (1994a) offers a similarly useful review of software for GLMs.

1.3.1 S-PLUS

S-PLUS is a general purpose statistical package available from Insightful Corporation. See http://www.insightful.com for information and pricing.

S-PLUS is the commercial version of S code originally developed by research statisticians at AT&T. The S language became popular in academic circles to which it was freely distributed and supported. Once commercialized, S-PLUS maintained its popularity with academicians. Gradually S-PLUS found favor with corporate research institutions and is currently one of the most popular statistical packages in use.

Behind GLIM, S-PLUS was one of the first packages to implement an algorithm for generalized linear models, or GLMs. First authored by Trevor Hastie, the GLM procedure has remained virtually unchanged since its inception. It uses a basic IRLS algorithm with an expected information matrix used to derive standard errors. S-PLUS supports the traditional GLM families, but does not estimate geometric or negative binomial models.

S-PLUS has no built-in support for fitting GEE models. However, there are various user-written macros or packages available on the Internet for fitting these and other similarly oriented models. S-PLUS is in wide use by researchers due to its feature-rich collection of commands, object-oriented design, and support for linking directly to externally compiled code. The original user-written software for fitting PA-GEE models was a program written in the C language. This program is accessed directly from within the S-PLUS system.

Other user-written programs are available including YAGS (yet another GEE solver). This package was originally developed for use with S-PLUS version 3.4, which has since been superseded by a number of version enhancements. At this writing the current version is S-PLUS 6.0. Since the methods for accessing externally compiled code have changed through the versions of S-PLUS, the YAGS package is not usable with the current version of S-PLUS. A community-based programming language founded on the syntax of S-PLUS called R may be substituted in many analyses; the current version of YAGS can also be used with R. The R language is fast becoming a popular means of performing statistical analysis within the academic community, much like S was in the 1980s.

geex is another GEE package for use with S-PLUS. It is available from http://lib.stat.cmu.edu/S/geex. While this package is somewhat older than currently developed packages, the geex software can be used with current versions of S-PLUS and is the GEE module we use for many examples in this text.

Graphics and support for low-level control of graphics are a particularly strong feature of the package. Nearly all of the graphics in this text were produced using S-PLUS.

1.3.2 SAS

SAS is a general purpose statistical package with the largest user base of any statistical software product. Information and pricing may be obtained from the SAS Institute company website at http://www.sas.com.

A user defined SAS macro was the only recognized program for estimating GEE models in the late 1980s. The macro estimated PA-GEE models for the Gaussian, binomial, Poisson, and gamma families. It was used exclusively until other packages, including SAS, implemented the routines into their main statistical offerings.

SAS incorporates a full range of PA-GEE model estimation as part of its STAT/GENMOD built-in procedure. GENMOD is an all-purpose GLM modeling facility that was written by Gordon Johnston in the early 1990s. User-written facilities were brought together from sources such as Hilbe (1994b) to complete this procedure for GLMs. In the mid-1990s a REPEATED option was added to the main GENMOD program. The new option allows estimation of the standard PA-GEE models using either moment estimators or alternating logistic regressions. It also allows estimation of the dispersion parameter via a user specified option for two different moment estimators. It is possible to perform GEE estimation on all standard GLM families.

There are a number of user-written macro additions that may be obtained and used as alternatives to those commands included in the software as well as additions to the base package.

In addition, SAS has excellent features for fitting likelihood-based models for panel data. Two of the highlights of this support include mixed linear regression models and mixed nonlinear regression models.

1.3.3 Stata

Stata is an all purpose statistical package that has excellent support for a variety of panel models. Stata's GEE program, called xtgee, is built upon its glm program. However, unlike SAS, Stata's GEE program is a separate command. Hilbe wrote the first comprehensive glm program for Stata in 1993. Hardin wrote the xtgee code several years afterwards. Hilbe (1993a) wrote the original glm user-written command, and the current Stata version of glm was written by Hardin and Hilbe in 2000.

Stata has good support for PA-GEE models along with options for specifying two different estimators for the dispersion parameter. All GLM families are included as modeling options, including power and odds power links.

In addition to its basic xtgee program, Stata employs separate program procedures for specific models. For instance, the command xtpois can be used to estimate either population averaged (PA-GEE), conditional fixed effects, or random effects Poisson models. The random effect may in turn be designated as belonging to the gamma or Gaussian distribution. This facility allows the user to easily compare alternative models.

In addition to the built-in support for PA-GEE models, Stata users have contributed programs for fitting generalized linear and latent models that include multi-level random effects models. This addition includes an adaptive quadrature optimization routine that outperforms the usual (nonadaptive) Gauss–Hermite quadrature implementations of many random-effects models. Stata also has a rather extensive suite of survey models, allowing the user to select strata and survey weights.

Information and pricing are available from http://www.stata.com.

1.3.4 SUDAAN

SUDAAN, developed by Research Triangle Institute in North Carolina, is a general purpose survey and panel data modeling package that can be used alone or as a callable program from within the SAS package. Information and pricing is available from http://www.rti.org/sudaan.

SUDAAN was specifically designed to analyze complex survey data. However, it can also be used for data without a specified sampling plan. While SUDAAN does not allow the modeling of several mainstream GLM families and links, it does include certain features not found in the other packages. Most notable among these is support for PA-GEE multinomial logistic regression models (demonstrated in this text). All SUDAAN commands include support for specification of strata and survey weights.

SUDAAN provides an option to estimate standard errors for its PA-GEE models using jackknife procedures. Stata allows this option, as well as bootstrapped standard errors, for its `glm` command, but it is not as yet implemented for the `xtgee` command.

Bieler and Williams (1997) have written an excellent user manual for SUDAAN that provides numerous examples of how to use and interpret GEE models using SUDAAN software.

1.4 Exercises

1. Obtain the documentation for the GLM and GEE commands that are supported by the software packages that you will use.

2. The table of link functions does not include entries for the negative binomial distribution (for specified ancillary parameter). Derive the entries for this function.

3. Identify and discuss two methods for dealing with overdispersion in a GLM.

4. Construct a GLM IRLS algorithm specifically for the log-linked Poisson regression model using the programming language of your software of choice.

5. Create a list of GLM and GEE capabilities for each software package that you use. Include in your list the available link functions, variance functions, dispersion statistics, variance estimators, and other options.

6. Show how the link and variance functions can be abstracted from the exponential family form of the GLM probability distribution.

CHAPTER 2

Model Construction and Estimating Equations

In this chapter we review various modeling techniques in order to provide a common glossary of terms to be used throughout the text. This review also provides a valuable base to which one can refer when generalizations are introduced.

We begin with a review of likelihood-based regression models. Our discussion of the standard techniques for deriving useful models begins with illustrations on independent data. Our focus is on the derivation of the likelihood and estimating equation. After illustrating the standard techniques for building estimating equations for likelihood-based models, we review the estimating equation for generalized linear models (GLMs). We point out the generalizations and relationship of GLMs to the probability-based models that precede them in the discussion.

After reviewing the techniques for model construction with independent data, we introduce the concepts associated with panel data and highlight the likelihood-based techniques for addressing second-order dependence within the data. Finally, we present estimators for the variance of the regression coefficients so that similar estimators for generalized estimating equations may be subsequently introduced with context.

2.1 Independent data

A common introduction to likelihood-based model construction involves several standard steps which follow:

1. Choose a distribution for the outcome variable.
2. Write the joint distribution for the data set.
3. Convert the joint distribution to a likelihood.
4. Generalize the likelihood via introduction of a linear combination of covariates and associated coefficients.
5. Parameterize the linear combination of covariates to enforce range restrictions on the mean and variance implied by the distribution.
6. Write the estimating equation for the solution of unknown parameters.

Once the model is derived, we may choose to estimate the fully specified log-likelihood with any extra parameters, or we may consider those extra parameters ancillary to the analysis. The former is called full information maximum likelihood (FIML); the latter is called limited information maximum likelihood (LIML). Estimation may then be carried out using an optimization method. The most common technique is that of Newton–Raphson, or a modification of the Newton–Raphson estimating algorithm.

We present an overview of this and other optimization techniques in section 2.4. For a detailed derivation of the Newton–Raphson and iteratively reweighted least squares (IRLS) algorithms, see Hardin and Hilbe (2001). For a practical discussion of optimization in general, see Gill, Murray, and Wright (1981).

The next three subsections illustrate model construction for three specific distributions. In carrying out the derivation of each respective model, we emphasize the steps to model construction, the need for parameterization, and the identification of the estimating equation.

2.1.1 The FIML estimating equation for linear regression

Let us assume that we have a data set where the outcome variable of interest is (effectively) continuous with a large range. In this situation the normal (Gaussian) distribution is typically used as the foundation for estimation. The density for the normal distribution $N(\mu, \sigma^2)$ is expressed as:

$$f(y|\mu, \sigma^2) = \frac{1}{\sqrt{2\pi\sigma^2}} \exp\left\{-\frac{(y-\mu)^2}{2\sigma^2}\right\} \tag{2.1}$$

where

$$\mathrm{E}(y) = \mu \in \Re \tag{2.2}$$
$$\mathrm{V}(y) = \sigma^2 > 0 \tag{2.3}$$

and $\Re$ indicates the range of real numbers. We may write this density for a single outcome as

$$f(y_i|\mu, \sigma^2) = \frac{1}{\sqrt{2\pi\sigma^2}} \exp\left\{-\frac{(y_i-\mu)^2}{2\sigma^2}\right\} \tag{2.4}$$

The joint density for n independent outcomes subscripted from $1, \ldots, n$ is the product of the densities for the individual outcomes

$$f(y_1, \ldots, y_n|\mu, \sigma^2) = \prod_{i=1}^{n} \frac{1}{\sqrt{2\pi\sigma^2}} \exp\left\{-\frac{(y_i-\mu)^2}{2\sigma^2}\right\} \tag{2.5}$$
$$= \prod_{i=1}^{n} \exp\left\{-\frac{1}{2}\ln\left(2\pi\sigma^2\right) - \frac{(y_i-\mu)^2}{2\sigma^2}\right\} \tag{2.6}$$

The likelihood is simply a restatement of the joint density where we consider

INDEPENDENT DATA

the outcomes as given, and model the parameters as unknown

$$L(\mu, \sigma^2 | y_1, \ldots, y_n) = \prod_{i=1}^{n} \exp\left\{-\frac{1}{2}\ln(2\pi\sigma^2) - \frac{(y_i - \mu)^2}{2\sigma^2}\right\} \quad (2.7)$$

$$= \exp\left\{\sum_{i=1}^{n}\left[-\frac{1}{2}\ln(2\pi\sigma^2) - \frac{(y_i - \mu)^2}{2\sigma^2}\right]\right\} \quad (2.8)$$

Since our goal is to introduce covariates that model the outcome, we add a subscript to the notation, changing μ to μ_i, allowing the mean to reflect a dependence on a linear combination of the covariates and their associated coefficients.

$$L(\boldsymbol{\mu}, \sigma^2 | y_1, \ldots, y_n) = \prod_{i=1}^{n} \exp\left\{-\frac{1}{2}\ln(2\pi\sigma^2) - \frac{(y_i - \mu_i)^2}{2\sigma^2}\right\} \quad (2.9)$$

$$= \exp\left\{\sum_{i=1}^{n}\left[-\frac{1}{2}\ln(2\pi\sigma^2) - \frac{(y_i - \mu_i)^2}{2\sigma^2}\right]\right\} \quad (2.10)$$

We introduce covariates into the model as a function of the expected value of the outcome variable. We also assume that we have a collection of independent covariates with associated coefficients to be estimated. The linear combination of the covariates and the associated coefficients is called the *linear predictor*, $\eta_i = \mathbf{x}_i \boldsymbol{\beta} \in \Re$, where $\mathbf{x}_i$ is the ith row of the $\mathbf{X}$ matrix. The linear predictor is introduced into the model in such a way that the range restrictions of the distribution are observed.

For this particular case, the variance of the outcome is $V(y_i) = \sigma^2$, which does not impose any range restrictions on the expected value of the outcome. Further, the range of the expected value matches the range of the linear predictor. As such, we could simply replace the expected value μ with the linear predictor. Formally, we use the identity function to parameterize the mean as

$$g(\mu_i) = \mu_i = \mathbf{x}_i \boldsymbol{\beta} \quad (2.11)$$

Under this approach, equation 2.12 is our likelihood-based model for linear regression. Replacing the expected value with our suitably parameterized linear predictor results in the log-likelihood

$$\mathcal{L}(\boldsymbol{\beta}, \sigma^2 | \mathbf{X}, y_1, \ldots, y_n) = \sum_{i=1}^{n}\left\{-\frac{1}{2}\ln(2\pi\sigma^2) - \frac{(y_i - \mathbf{x}_i\boldsymbol{\beta})^2}{2\sigma^2}\right\} \quad (2.12)$$

Even though the identity parameterization is the natural, canonical, parameterization from the derivation, we are not limited to that choice. In the case that the outcomes are always positive, we could choose

$$g(\mu_i) = \ln(\mu_i) = \mathbf{x}_i \boldsymbol{\beta} \quad (2.13)$$

resulting in the familiar log-linear regression model. The parameterization via the log function implies that $g^{-1}(\mathbf{x}_i\boldsymbol{\beta}) = \exp(\mathbf{x}_i\boldsymbol{\beta}) = \mu_i$, and ensures a desired nonnegative fit from the linear predictor. Under this log parameterization, our final log-likelihood model for log-linear regression is

$$\mathcal{L}(\beta, \sigma^2 | \mathbf{X}, y_1, \ldots, y_n) = \sum_{i=1}^{n} \left\{ -\frac{1}{2} \ln\left(2\pi\sigma^2\right) - \frac{(y_i - \exp(\mathbf{x}_i\beta))^2}{2\sigma^2} \right\} \quad (2.14)$$

For a likelihood-based model, the next step is to specify the *estimating equation*. The solution to the estimating equation provides the desired estimates. In the case of a likelihood-based model, the estimating equation is the derivative of the log-likelihood. We either derive an estimating equation in terms of $\Theta = (\beta, \sigma^2)$ (a FIML model), or we specify an estimating equation in terms of $\Theta = (\beta)$ where σ^2 is ancillary (a LIML model). The ancillary parameters in a LIML model are either estimated separately, or specified. The resulting estimates for the parameters are conditional on the ancillary parameters being correct.

Using the identity link for parameterization of the linear predictor, the linear regression FIML estimating equation $\Psi(\Theta) = \mathbf{0}$ for $(\beta_{p\times 1}, \sigma^2)$ is given by

$$\begin{bmatrix} \left\{ \dfrac{\partial \mathcal{L}}{\partial \beta_j} = \sum_{i=1}^{n} x_{ji} \dfrac{1}{\sigma^2}(y_i - \mathbf{x}_i\beta) \right\}_{j=1,\ldots,p} \\ \dfrac{\partial \mathcal{L}}{\partial \sigma^2} = -\sum_{i=1}^{n} \left(\dfrac{1}{2\sigma^2} + \dfrac{(y_i - \mathbf{x}_i\beta)^2}{2\sigma^4} \right) \end{bmatrix}_{(p+1)\times 1} = [\mathbf{0}]_{(p+1)\times 1} \quad (2.15)$$

Note, however, that we write the estimating equation in terms of μ, rather than $\mathbf{x}\beta$, to incorporate a general parameterization of the linear predictor. To include the parameterization, we use the chain rule

$$\frac{\partial \mathcal{L}}{\partial \beta} = \frac{\partial \mathcal{L}}{\partial \mu} \frac{\partial \mu}{\partial \eta} \frac{\partial \eta}{\partial \beta} \quad (2.16)$$

In this more general notation, the estimating equation $\Psi(\Theta) = \mathbf{0}$ is given by

$$\begin{bmatrix} \left\{ \dfrac{\partial \mathcal{L}}{\partial \beta_j} = \sum_{i=1}^{n} \dfrac{1}{\sigma^2}(y_i - \mu_i)\left(\dfrac{\partial \mu}{\partial \eta}\right)_i x_{ji} \right\}_{j=1,\ldots,p} \\ \dfrac{\partial \mathcal{L}}{\partial \sigma^2} = -\sum_{i=1}^{n} \left(\dfrac{1}{2\sigma^2} + \dfrac{(y_i - \mu_i)^2}{2\sigma^4} \right) \end{bmatrix}_{(p+1)\times 1} = [\mathbf{0}]_{(p+1)\times 1}$$

(2.17)

and we must specify the relationship (parameterization) of the expected value μ to the linear predictor $\eta = \mathbf{X}\beta$. In the case of linear regression, $\eta = \mu$.

The estimating equation for the LIML model $\Psi[\Theta = (\beta)] = \mathbf{0}$, treating σ^2 as ancillary, is just the upper $p \times 1$ part of the estimating equation 2.17.

INDEPENDENT DATA

2.1.2 The FIML estimating equation for Poisson regression

The Poisson distribution is the natural choice to model outcome variables that are nonnegative counts. The Poisson density is given by

$$f(y|\lambda) = \frac{e^{-\lambda}\lambda^y}{y!} \tag{2.18}$$

where

$$E(y) = \lambda > 0 \tag{2.19}$$
$$V(y) = \lambda > 0 \tag{2.20}$$

The joint density for n independent outcomes subscripted from $1,\ldots,n$ is then given as the product of the densities for the individual outcomes

$$f(y_1,\ldots,y_n|\lambda) = \prod_{i=1}^{n} \frac{e^{-\lambda}\lambda^{y_i}}{y_i!} \tag{2.21}$$

$$= \prod_{i=1}^{n} \exp\{-\lambda + y_i \ln(\lambda) - \ln(y_i!)\} \tag{2.22}$$

$$= \prod_{i=1}^{n} \exp\{-\lambda + y_i \ln(\lambda) - \ln \Gamma(y_i+1)\} \tag{2.23}$$

The likelihood is a restatement of the joint density where we consider the outcomes as given and model the parameter as unknown

$$L(\lambda|y_1,\ldots,y_n) = \prod_{i=1}^{n} \exp\{-\lambda + y_i \ln(\lambda) - \ln \Gamma(y_i+1)\} \tag{2.24}$$

Since our goal is to introduce covariates that model the outcome, we add a subscript to the notation allowing the mean to reflect a dependence on a linear combination of the covariates and their associated coefficients. We also replace the usual presentation of the Poisson distribution using μ for the expected value λ. Replacing λ with μ is merely for notational consistency (with the models to follow), and has no effect on the derivation of the estimating equation.

$$L(\mu|y_1,\ldots,y_n) = \prod_{i=1}^{n} \exp\{-\mu_i + y_i \ln(\mu_i) - \ln \Gamma(y_i+1)\} \tag{2.25}$$

As in the previous derivation for linear regression, we introduce covariates into the model through the expected value μ of the outcome variable, and we assume a collection of independent covariates with associated coefficients to be estimated called the linear predictor $\eta_i = \mathbf{x}_i\beta \in \Re$; note that $\mathbf{x}_i$ is the ith row of the design matrix $\mathbf{X}$.

We introduce the linear predictor into the model in such a way that the range restrictions of the distribution are observed. In this particular case, the variance of the outcome is given by

$$V(y_i) = \mu_i > 0 \qquad (2.26)$$

which depends on the expected value of the outcome. In fact, for the Poisson distribution, the variance is equal to the expected value. Therefore, we should parameterize the linear predictor to enforce a range $(0, \infty)$. The natural, or canonical, choice obtained from our derivation is

$$g(\mu_i) = \ln(\mu_i) = \mathbf{x}_i \boldsymbol{\beta} \qquad (2.27)$$

This parameterization of the canonical Poisson link function implies an inverse relationship given by $g^{-1}(\mathbf{x}_i \boldsymbol{\beta}) = \exp(\mathbf{x}_i \boldsymbol{\beta}) = \mu_i$, which ensures a nonnegative fit from the linear predictor. Under this parameterization for the expected value, the final log-likelihood is given by

$$\mathcal{L}(\boldsymbol{\beta}|\mathbf{X}, y_1, \ldots, y_n) = \sum_{i=1}^{n} \left\{ -\exp(\mathbf{x}_i \boldsymbol{\beta}) + y_i \mathbf{x}_i \boldsymbol{\beta} - \ln \Gamma(y_i + 1) \right\} \qquad (2.28)$$

The general FIML estimating equation $\Psi(\Theta) = \mathbf{0}$ for $\Theta = (\boldsymbol{\beta})$ is then

$$\left[\left\{ \frac{\partial \mathcal{L}}{\partial \beta_j} = \sum_{i=1}^{n} \left(\frac{y_i}{\mu_i} - 1 \right) \left(\frac{\partial \mu}{\partial \eta} \right)_i x_{ji} \right\}_{j=1,\ldots,p} \right]_{p \times 1} = [\mathbf{0}]_{p \times 1} \qquad (2.29)$$

where there are no ancillary parameters.

2.1.3 The FIML estimating equation for Bernoulli regression

Assume that the outcome variable of interest is binary and that our data are coded such that a successful outcome in the experiment is coded as a one and a failure is coded as a zero. The Bernoulli distribution, a limiting case of the binomial distribution, is the appropriate choice for estimation of binary data. Its density function is

$$f(y|p) = p^y (1-p)^{1-y} \qquad (2.30)$$

where $p \in [0, 1]$ is the probability of success, and

$$\mathrm{E}(y) = p \in (0, 1) \qquad (2.31)$$
$$\mathrm{V}(y) = p(1-p) \in (0, 1) \qquad (2.32)$$

The joint density for n independent outcomes subscripted from $1, \ldots, n$ is then given as the product of the densities for the individual outcomes

$$f(y_1, \ldots, y_n|p) = \prod_{i=1}^{n} p^{y_i} (1-p)^{1-y_i} \qquad (2.33)$$

$$= \prod_{i=1}^{n} \exp \left\{ y_i \ln \left(\frac{p}{1-p} \right) + \ln(1-p) \right\} \qquad (2.34)$$

The likelihood is simply a restatement of the joint density where we consider the outcomes as given and the parameters are modeled as unknown.

$$L(p|y_1,\ldots,y_n) = \prod_{i=1}^{n} \exp\left\{y_i \ln\left(\frac{p}{1-p}\right) + \ln(1-p)\right\} \quad (2.35)$$

Since our goal is to introduce covariates that model the outcome, and since we are interested in the individual contributions of each subject to the model, we introduce a subscript to the notation. This notation, changing p to p_i, allows the mean response to reflect a dependence on the linear combination of the covariates and their associated coefficients. We also replace the common presentation of the Bernoulli expected value p with μ. In so doing, we have a consistent notation among various distributions.

$$L(\boldsymbol{\mu}|y_1,\ldots,y_n) = \prod_{i=1}^{n} \exp\left\{y_i \ln\left(\frac{\mu_i}{1-\mu_i}\right) + \ln(1-\mu_i)\right\} \quad (2.36)$$

Again, we introduce covariates into the model through the expected value of the outcome variable. As in the previous example, we assume a collection of independent covariates with associated coefficients to be estimated called the linear predictor $\eta_i = \mathbf{x}_i \boldsymbol{\beta} \in \Re$.

We introduce the linear predictor into the model in such a way that the range restrictions of the distribution and variance are observed. In this particular case, the variance of the outcome is given by

$$V(y_i) = \mu_i(1 - \mu_i) \quad (2.37)$$

where $\mu_i \in (0,1)$ depends on the expected value of the outcome. Therefore, we should parameterize the linear predictor to enforce a range $(0,1)$.

The binomial admits several interesting and useful parameterizations. If we parameterize using the natural, or canonical, form from the derivation of the estimating equation

$$g(\mu_i) = \ln\left(\frac{\mu_i}{1-\mu_i}\right) = \mathbf{x}_i \boldsymbol{\beta} \quad (2.38)$$

we have a logistic, or logit, regression model. If we parameterize using

$$g(\mu_i) = \Phi^{-1}(\mu_i) = \mathbf{x}_i \boldsymbol{\beta} \quad (2.39)$$

where $\Phi^{-1}()$ is the inverse of the cumulative normal distribution (the quantile function), we have a probit, or normit, regression model. If we parameterize using

$$g(\mu_i) = -\ln\{-\ln(\mu_i)\} = \mathbf{x}_i \boldsymbol{\beta} \quad (2.40)$$

we have a log-log, or gompit, regression model.

The general FIML estimating equation $\Psi(\Theta) = \mathbf{0}$ for $\Theta = (\boldsymbol{\beta})$ is given by

$$\left[\left\{\frac{\partial \mathcal{L}}{\partial \beta_j} = \sum_{i=1}^{n} \left(\frac{y_i}{\mu_i} - \frac{1-y_i}{1-\mu_i}\right) \left(\frac{\partial \mu}{\partial \eta}\right)_i x_{ji}\right\}_{j=1,\ldots,p}\right]_{p \times 1} = [\mathbf{0}]_{p \times 1} \quad (2.41)$$

There are no ancillary parameters.

2.1.4 The LIML estimating equation for GLMs

In the preceding sections we introduced a level of specification that is not normally present in the illustration of likelihood-based models. The extra specification was in terms of parameterizing the linear predictor. The reason we introduced this specification was to motivate its use in the more general setting of deriving models for an entire *family* of distributions.

We looked at three specific distributions. Here, we investigate the exponential family of distributions. The advantage is that the exponential family not only includes the three specific examples already presented, but also includes many other distributions.

As we discussed in Chapter 1, the theory of generalized linear models (GLMs) was introduced in Nelder and Wedderburn (1972). These authors showed an underlying unity to a class of regression models where the response variable was a member of the exponential family of probability distributions. Again, members of this family of distributions include the Gaussian or normal, Bernoulli, binomial, Poisson, gamma, inverse Gaussian, geometric, and negative binomial distributions; see Hilbe (1993b) and Hilbe (1994c) for more information on the negative binomial as a GLM.

We can proceed in a similar manner to our previous examples with the goal of deriving likelihood-based models for this family. The exponential family of distributions has a location parameter θ, a scale parameter $a(\phi)$, and a normalizing term $c(y, \phi)$ with probability density

$$f(y; \theta, \phi) = \exp\left\{\frac{y\theta - b(\theta)}{a(\phi)} + c(y, \phi)\right\} \tag{2.42}$$

where

$$\mathrm{E}(y) = b'(\theta) = \mu \tag{2.43}$$
$$\mathrm{V}(y) = b''(\theta)a(\phi) \tag{2.44}$$

The normalizing term is independent of θ and ensures that the density integrates to one. We have not yet listed any range restrictions. Instead, the range restrictions are addressed after the estimating equation has been constructed. The variance is a function of the expected value of the distribution and a function of the (possibly unknown) scale parameter $a(\phi)$. The density for a single observation is

$$f(y_i; \theta, \phi) = \exp\left\{\frac{y_i\theta - b(\theta)}{a(\phi)} + c(y_i, \phi)\right\} \tag{2.45}$$

and the joint density for a set of n independent outcomes subscripted from $1, \ldots, n$ is the product of the densities for the individual outcomes

$$f(y_1, \ldots, y_n; \theta, \phi) = \prod_{i=1}^{n} \exp\left\{\frac{y_i\theta - b(\theta)}{a(\phi)} + c(y_i, \phi)\right\} \tag{2.46}$$

INDEPENDENT DATA

The likelihood is simply a restatement of the joint density where we consider the outcomes as given, and model the parameters as unknown

$$L(\theta, \phi | y_1, \ldots, y_n) = \prod_{i=1}^{n} \exp\left\{\frac{y_i \theta - b(\theta)}{a(\phi)} + c(y_i, \phi)\right\} \quad (2.47)$$

Instead of introducing the covariates into the model at this point, it is notationally advantageous to instead wait and introduce covariates into the estimating equation. We now include a subscript for θ in anticipation of introducing the covariates.

The log-likelihood for the exponential family is

$$\mathcal{L}(\boldsymbol{\theta}, \phi | y_1, \ldots, y_n) = \sum_{i=1}^{n} \left\{\frac{y_i \theta_i - b(\theta_i)}{a(\phi)} + c(y_i, \phi)\right\} \quad (2.48)$$

The goal is to obtain a maximum likelihood estimator for θ. Since our focus is only on θ, we derive a LIML estimating equation where we treat the dispersion parameter $a(\phi)$ as ancillary.

We know from basic principles that

$$E\left(\frac{\partial \mathcal{L}}{\partial \theta}\right) = 0 \quad (2.49)$$

Our LIML estimating equation is then $\Psi(\Theta) = \partial \mathcal{L}/\partial \theta = \mathbf{0}$ where we derive

$$\frac{\partial \mathcal{L}}{\partial \theta} = \sum_{i=1}^{n} \frac{y_i - b'(\theta)}{a(\phi)} \quad (2.50)$$

Utilizing the GLM result that in canonical form $b'(\theta) = \mu$, we may write

$$\frac{\partial \mathcal{L}}{\partial \theta} = \sum_{i=1}^{n} \frac{y_i - \mu_i}{a(\phi)} \quad (2.51)$$

substituting our preferred (consistent) μ notation for the expected value.

Since our goal is to introduce covariates that model the outcome, we included a subscript on μ allowing the mean to reflect a dependence on a linear combination of the covariates and their associated coefficients. We can now use the chain rule to obtain a more useful form of the LIML estimating equation $\Psi(\Theta) = \mathbf{0}$ for $\Theta = \boldsymbol{\beta}_{(p \times 1)}$

$$\frac{\partial \mathcal{L}}{\partial \beta} = \left[\left(\frac{\partial \mathcal{L}}{\partial \theta}\right)\left(\frac{\partial \theta}{\partial \mu}\right)\left(\frac{\partial \mu}{\partial \eta}\right)\left(\frac{\partial \eta}{\partial \beta_j}\right)\right]_{(p \times 1)} \quad (2.52)$$

$$= \left[\sum_{i=1}^{n} \left(\frac{y_i - b'(\theta_i)}{a(\phi)}\right)\left(\frac{1}{V(\mu_i)}\right)\left(\frac{\partial \mu}{\partial \eta}\right)_i (x_{ji})\right]_{(p \times 1)} \quad (2.53)$$

$$= \left[\sum_{i=1}^{n} \frac{y_i - \mu_i}{a(\phi) V(\mu_i)} \left(\frac{\partial \mu}{\partial \eta}\right)_i x_{ji}\right]_{(p \times 1)} \quad (2.54)$$

so that the general LIML estimating equation for the exponential family is given by

$$\left[\left\{ \frac{\partial \mathcal{L}}{\partial \beta_j} = \sum_{i=1}^{n} \frac{y_i - \mu_i}{a(\phi) \mathrm{V}(\mu_i)} \left(\frac{\partial \mu}{\partial \eta} \right)_i x_{ji} \right\}_{j=1,\ldots,p} \right]_{p \times 1} = [0]_{p \times 1} \quad (2.55)$$

The result is an estimating equation derived from the exponential family of distributions where the expected value of the outcome variable is parameterized in terms of a linear predictor. The estimating equation is the derivative of the log-likelihood for the exponential family. It is given in terms of the suitably parameterized expected value and variance, where the variance is a function of the expected value. There is an additional parameter $a(\phi)$ that is not addressed in the estimating equation; it is an ancillary parameter called the *dispersion* in GLM literature.

Our presentation in this section assumes that μ and $\mathrm{V}(\mu)$ are the resulting forms from the chosen exponential family member distribution. Since the expected value and variance function result from the specific distribution in the exponential family, the estimating equation implies a valid likelihood (in terms of the source distribution of our mean and variance functions). Consequently, under these restrictions we view equation 2.55 as the LIML estimating equation for GLMs.

The ancillary parameter $a(\phi)$ is taken to be the scale parameter ϕ in nearly all GLM software implementations. One of the software implementations (used in examples in this text) of GEE-based extensions to GLMs, however, allows a more general setting. As such, and in anticipation of later explanation, our presentation leaves this ancillary parameter specified as $a(\phi)$.

We lastly turn to a discussion regarding the restriction of the range of our parameterized linear predictor. GLMs are specified through a parameterization function, called the *link function*, and a variance that is a function of the mean. The conservative approach is to specify only parameterizations that ensure implied range restrictions of the mean and variance functions. In so doing, the optimization should have no numeric difficulties (outside of collinearity or poorly chosen starting values) iterating to the global solution. However, if we choose a link function that does not restrict the variance to positive solutions, optimization may step to a candidate solution for which the variance is negative or undefined. For example, if we choose the log-link for a binomial variance model, the calculation of $\exp(\mathbf{x}_i \beta)$ might be larger than one or smaller than zero for certain observations. On the other hand, the data might support this link in the sense that the calculation of $\exp(\mathbf{x}_i \beta) \in (0, 1)$ for all i for the true β. If the data support a nonrestrictive link, then we are free to fit the model with this (nonrestrictive) link function and inference is clear. An example of the application and interpretation of nonrestrictive links is given in Wacholder (1986). While data may occasionally support a nonrestrictive link function, we are not surprised when unrestricted optimization steps out of the restricted range implied by the variance function. In other words, we can, in fact, use any link function. Whether estimation proceeds

INDEPENDENT DATA

to a solution in a valid region of the parameter space using particular data, however, is not guaranteed.

2.1.5 The LIMQL estimating equation for GLMs

In the first three examples of model construction in this chapter, we introduced covariates into the log-likelihood of a model and then derived the associated estimating equation. Our presentation of likelihood-based generalized linear models, however, instead introduced the covariates directly into the estimating equation. Either way, the result was an estimating equation that included a linear predictor with an associated coefficient vector to be estimated.

A powerful result from Wedderburn (1974) allows us to view the mean and variance functions as part of the LIML estimating equation for GLMs with no formal restriction that they originate from a specific (or even the same) distribution. If we choose μ and $V(\mu)$ from a (single) member-distribution of the exponential family, the estimating equation then implies the associated log-likelihood for that distribution. Resulting coefficient estimates in this case are properly labeled maximum likelihood estimates.

Wedderburn's work extends the result by assuming that the form of the variance function is a known function of the mean (up to a scalar constant) and by assuming independence of the observations. These are weaker assumptions than a derivation from a specific distribution. This extension of work under a weaker set of assumptions is analogous to Gauss's extension of classical ordinary least squares where the properties of the estimates for linear regression are justified on assumptions of independence and constant variance rather than upon an assumption of normality. We are therefore free to choose any parameterization of the mean and variance function and apply them in the derived estimating equation.

When we choose functions that are not from an exponential family member, the log-likelihood implied by the estimating equation is called a *quasilikelihood* defined as

$$\mathcal{Q}(y; \mu) = \int^{\mu} \frac{y - \mu^*}{V(\mu^*)a(\phi)} d\mu^* \tag{2.56}$$

Resulting coefficient estimates are properly called maximum quasilikelihood estimates. The quasilikelihood is a generalization of the likelihood. Often, one refers to all estimates obtained from a GLM as maximum quasilikelihood estimates, irrespective of the source distribution of the applied mean and variance functions. This is technically the case for all models except those employing the canonical link, which, in fact, do produce likelihood-based estimates.

We need not alter the LIML estimating equation given in the preceding section. The LIMQL estimating equation for GLMs with no restriction on the choice of the mean and variance functions is the same (equation 2.55) as the case where we restricted the population of candidate choices for the mean and variance functions.

2.2 Estimating the variance of the estimates

Included in the original presentation of GLMs was a description of an iteratively reweighted least squared (IRLS) algorithm for fitting the models and obtaining estimates. This algorithm is iterative and requires only weighted OLS at each step. The majority of programmable statistical software packages can be programmed to implement the full collection of models. A presentation of the details and derivation of the IRLS algorithm together with estimated variance matrices is covered in Hardin and Hilbe (2001). Here, we discuss the results derived from that reference.

We present formulae and a short discussion on various estimated variance matrices where notationally $\partial \mu / \partial \eta$ is to be calculated at $\mu = \hat{\mu}$ and $\hat{\phi}$ is an estimate of the dispersion parameter $a(\phi)$. Full details on GLMs and their associated variance estimates are in the references cited. Additional coverage of GLMs can be found in McCullagh and Nelder (1989), Hilbe (1994a), and Lindsey (1997).

Statistical packages typically calculate the variance estimate numerically, or analytically, as the inverse matrix of (negative) second derivatives. Alternatively, the estimate may be constructed from the Fisher scoring matrix of expected second derivatives. In the case that the GLM is fit with the canonical link, these calculations result in the same estimate. Otherwise, the two estimates are only asymptotically the same.

The variance estimates are given by

$$\widehat{\mathbf{V}}_{\mathrm{H}}(\hat{\boldsymbol{\beta}}) = \left\{ \left(-\frac{\partial^2 \mathcal{L}}{\partial \beta_u \partial \beta_v} \right) \right\}^{-1}_{p \times p} \tag{2.57}$$

where $u, v = 1, \ldots, p$, and p is the column dimension of $\mathbf{X}$. The Hessian matrix uses the second derivatives (of the likelihood) while the Fisher scoring matrix uses the expected second derivatives. If the second derivatives are used, we illustrate this by denoting $\widehat{\mathbf{V}}_{\mathrm{H}}$ as $\widehat{\mathbf{V}}_{\mathrm{OH}}$ to indicate that the variance estimate is based on the observed Hessian. If the Fisher scoring matrix is used, we denote $\widehat{\mathbf{V}}_{\mathrm{H}}$ as $\widehat{\mathbf{V}}_{\mathrm{EH}}$ to indicate that the variance estimate is based on the expected Hessian. The approaches are based on two different (asymptotically equivalent) forms of the information matrix.

The sandwich estimate of variance is of the form $\mathbf{A}^{-1} \mathbf{B} \mathbf{A}^{-\mathrm{T}}$ where $\mathbf{A}$ is the usual estimate of the variance based on the information matrix. The middle of the sandwich is a correction term. Huber (1967) introduced the idea in a discussion of the general properties inherent in the solution of an estimating equation given by

$$\left[\Psi = \sum_{i=1}^{n} \Psi_i(\mathbf{x}_i, \boldsymbol{\beta}) \right]_{p \times 1} = [\mathbf{0}]_{p \times 1} \tag{2.58}$$

where $\Psi_i(\mathbf{x}_i, \boldsymbol{\beta})$ is the estimating equation for the ith observation.

For our likelihood-based models, the estimating equation is the derivative of the log-likelihood for the distribution, $\partial \mathcal{L} / \partial \boldsymbol{\beta}$. Our desire is to evaluate the properties of the variance estimate for cases when the data really come from distribution g.

$\mathbf{A}$ is formally given by

$$\mathbf{A} = \left(\frac{\partial \mathrm{E}\Psi(\beta)}{\partial \beta}\right)^{-1} \quad (2.59)$$

In most cases we can swap the order of the expectation and derivative operators so that

$$\mathbf{A} = \mathrm{E}\left(\frac{\partial \Psi(\beta)}{\partial \beta}\right)^{-1} \quad (2.60)$$

may be estimated as (the inverse of) the usual estimate of variance based on the information matrix $\widehat{\mathbf{V}}_H$—a naive variance estimate assuming the data are from distribution f. Cases which allow swapping of the order of expectation and differentiation are validated through convergence theorems not covered in this text. Interested readers should look at a text which more formally covers regularity conditions, such as Billingsley (1986), for details. Otherwise, note that the interchange of these operators is allowed in the various models we discuss.

The correction term given by the $\mathbf{B}$ matrix is the covariance of the estimating equation $\Psi(\beta) = \sum \Psi_i(\mathbf{x}_i, \beta)$; it is the covariance matrix of a sum (of vectors). Since the expected value of the score contributions is zero, the variance of the estimating equation is simply $\mathrm{E}(\Psi(\beta)^\mathrm{T}\Psi(\beta))$ so that

$$\mathbf{B} = \sum_{i=1}^{n} \mathrm{E}\left[\Psi_i(\mathbf{x}_i,\beta)\Psi_i^\mathrm{T}(\mathbf{x}_i,\beta)\right] + \sum_{i=1}^{n}\sum_{\substack{j=1\\j\neq i}}^{n} \mathrm{E}\left[\Psi_i(\mathbf{x}_i,\beta)\Psi_i^\mathrm{T}(\mathbf{x}_j,\beta)\right] \quad (2.61)$$

If we assume that the observations are independent, then the cross terms are zero and the natural estimator of $\mathbf{B}$ is

$$\widehat{\mathbf{B}} = \sum_{i=1}^{n}\left[\widehat{\Psi}_i(\mathbf{x}_i,\widehat{\beta})\widehat{\Psi}_i^\mathrm{T}(\mathbf{x}_i,\widehat{\beta})\right] \quad (2.62)$$

Using this information, the middle of the sandwich variance estimate is formed from the independent contributions of the estimating equation. For example, the correction term for the sandwich estimate of variance appropriate for GLMs is derived using

$$\widehat{\Psi}_i(\mathbf{x}_i,\widehat{\beta}) = \left(\frac{\partial \mathcal{L}}{\partial \eta}\right)_i = \left(\frac{\partial \mathcal{L}}{\partial \mu}\right)_i\left(\frac{\partial \mu}{\partial \eta}\right)_i = \mathbf{x}_i^\mathrm{T}\frac{y_i-\widehat{\mu}_i}{\mathrm{V}(\widehat{\mu})_i}\left(\frac{\partial \mu}{\partial \eta}\right)_i \quad (2.63)$$

where the derivatives are evaluated at the estimated parameters and the correction term is given by

$$\widehat{\mathbf{B}}_{\mathrm{GLM}}(\widehat{\beta}) = \left[\sum_{i=1}^{n}\mathbf{x}_i^\mathrm{T}\left\{\frac{y_i-\widehat{\mu}_i}{\mathrm{V}(\widehat{\mu}_i)}\left(\frac{\partial \mu}{\partial \eta}\right)_i\widehat{\phi}\right\}^2\mathbf{x}_i\right]_{p\times p} \quad (2.64)$$

Included in this correction term, $\mathbf{x}_i$ is the ith $(p \times 1)$ row vector of the $(n \times p)$ matrix of covariates $\mathbf{X}$.

The general sandwich estimate of variance is the $p \times p$ matrix

$$\widehat{\mathbf{V}}_S(\widehat{\boldsymbol{\beta}}) = \widehat{\mathbf{V}}_H^{-1}(\widehat{\boldsymbol{\beta}})\widehat{\mathbf{B}}(\widehat{\boldsymbol{\beta}})\widehat{\mathbf{V}}_H^{-1}(\widehat{\boldsymbol{\beta}}) \tag{2.65}$$

Since the sandwich estimate of variance combines the variance estimate for the specified model with a variance matrix constructed from the data, the variance estimate is sometimes called the empirical variance estimate.

Note that we can write equation 2.63 as the product of $\mathbf{x}_i$ and some scalar quantity u_i. By construction, the expected value of u_i is zero. These individual values are called the *scores* or *score residuals*. Some software packages allow access to the scores for model assessment.

If observations may be grouped due to some correlation structure (perhaps because the data are really panel data), then the sandwich estimate is modified to consider the sums of the n_i observations for each independent panel i. The individual observation-level contributions to the estimating equation are no longer independent; but the sums over the panel of the contributions are independent. These contributions to the estimated scores are used to form the middle of the *modified sandwich variance*. Continuing our example for GLMs, the correction term is given by

$$\widehat{\mathbf{B}}_{MS}(\widehat{\boldsymbol{\beta}}) = \left[\sum_{i=1}^{n} \left\{ \sum_{t=1}^{n_i} \mathbf{x}_{it}^T \frac{y_{it} - \widehat{\mu}_{it}}{V(\widehat{\mu}_{it})} \left(\frac{\partial \mu}{\partial \eta}\right)_{it} \widehat{\phi} \right\}_{p \times 1} \left\{ \sum_{j=1}^{n_i} \frac{y_{it} - \widehat{\mu}_{it}}{V(\widehat{\mu}_{it})} \left(\frac{\partial \mu}{\partial \eta}\right)_{it} \widehat{\phi} \, \mathbf{x}_{it} \right\}_{1 \times p} \right]_{p \times p} \tag{2.66}$$

where $\mathbf{x}_{it}$ is the itth $(p \times 1)$ row vector of the $(n \times p)$ matrix of covariates $\mathbf{X}$ and $j = 1, \ldots, p$.

Using either form of the naive pooled variance estimate $\widehat{\mathbf{V}}_H$, which ignored any within-panel correlation, the modified sandwich estimate of variance is the $p \times p$ matrix given by

$$\widehat{\mathbf{V}}_{MS} = \widehat{\mathbf{V}}_H^{-1}(\widehat{\boldsymbol{\beta}})\widehat{\mathbf{B}}_{MS}(\widehat{\boldsymbol{\beta}})\widehat{\mathbf{V}}_H^{-1}(\widehat{\boldsymbol{\beta}}) \tag{2.67}$$

A sandwich estimate of variance constructed with $\mathbf{V}_H = \mathbf{V}_{OH}$ is called the *robust* variance estimate. If the construction uses $\mathbf{V}_H = \mathbf{V}_{EH}$, the variance estimate is called the *semi-robust* variance estimate. The distinction arises when the estimated semi-robust variance, constructed with the expected Hessian, is not robust to misspecification of the link function.

Again, note that we can write the relevant terms in the innermost sums of equation 2.66 as the product of $\mathbf{x}_{it}$ and some scalar quantity u_{it}. As in the case of uncorrelated data, the expected value of u_{it} is zero by construction. The u_{it} values are the scores.

We emphasize that the middle of the modified sandwich estimate of variance has replaced panels of observations with their respective sums. The rank of the resulting matrix is less than the number of panels in the construction. Therefore, this variance estimate should not be used for data sets with a small number of panels since the asymptotic justification of the distribution includes the assumption that the number of panels goes to infinity. This dependence on the number of panels can be seen on inspection of equation 2.64 where the result is computed by summing n matrices of size $(p \times p)$. Assuming an organization of these n observations in a balanced data set of k panels each with t observations $(n = kt)$, equation 2.66 is the result of summing only k matrices of size $(p \times p)$. If $k < p$, the modified sandwich estimate of variance is singular.

As a generalization, if the observations are not independent, but may be pooled into independent panels, the formation of the **B** matrix is a simple extension of the usual approach. The correlation within independent panels is addressed by summing the contributions to the estimating equation. The **B** matrix is the sum (over panels) of the outer product of the independent sums (within panels) of the estimating equation $\sum_i (\sum_t \Psi_{it})(\sum_t \Psi_{it})^{\mathrm{T}}$. In this case, the estimate of variance is called the modified sandwich variance estimate. The estimation is more difficult if the equation is not defined by independent observations nor by independent panels. For example, in the Cox proportional hazard model, observations contribute to the estimating equation through membership in the risk pool. Further, the risk pools share observations. Thus, the estimating equation is not characterized by either independent observations nor by independent panels of observations. Moreover, the derivation of the sandwich estimate of variance is complicated by the need to identify a suitable form of the estimating equation. Lin and Wei (1989) demonstrate the derivation of the sandwich estimate of variance for this complicated model.

Several times throughout the text we construct these modified sandwich estimates of variance for generalized estimating equations. The preceding discussion of sandwich estimates of variance is valid for the estimating equations derived from likelihoods as well as for the estimating equations that imply quasilikelihoods.

Lee, Scott, and Soo (1993) show that the modified sandwich estimate of variance for the pooled estimator underestimates the true covariance matrix. This is well known, and in fact all maximum likelihood estimation procedures underestimate the true covariance matrix. For small samples, this bias is more pronounced, and various ad hoc measures have been proposed for modifying the sandwich estimate of variance.

The most common modification to the sandwich estimate of variance is a scale factor that depends on the sample size. For the usual sandwich estimate of variance a commonly used approach multiplies the estimate by $n/(n-p)$ where n, in this case, is the sample size, and p is the number of covariates in the model. For the modified sandwich estimate of variance, the estimate is scaled by $n/(n-1)$ where n, in this case, is the number of panels. This ad hoc attempt to modify the sandwich estimate of variance for use with small samples results in different answers from competing software packages. The user should consult the documentation for specific software to learn if any scale adjustments are made.

2.3 Panel data

There is substantial literature addressing the subject of clustered data. Clustered data occur when there is a natural classification to observations such that data may be organized according to generation or sampling from units. For example, we may collect data on loans where we have multiple observations from different banks. It would be natural to address the dependence in the data on the bank itself, and there are several methods that we might utilize to take this dependence into account.

Panel data comes under many headings. If the panels represent a level of data organization where the observations within panels come from different experimental units belonging to the same classification, the data are usually called panel data, clustered data, or repeated measurement data. If the observations within panels come from the same experimental unit measured over time, the data are typically called longitudinal data.

Unless a model or method is specific to a certain type of panel structure, we adopt the term panel data to imply all forms of this type of data. Each method for addressing the panel structure of the data has advantages as well as limitations; it benefits the researcher to recognize the assumptions and inferences that are available.

In a panel data set, we assume that we have $i = 1, \ldots, n$ panels (clusters) where each panel has $t = 1, \ldots, n_i$ correlated observations. This notation allows either balanced panels, $n_1 = n_2 = \cdots = n_n$, or unbalanced panels $n_j \neq n_i$ for at least one $j \in \{1, \ldots, n\}, j \neq i$. We focus on the exponential family of distributions since they include the distributions individually illustrated in previous subsections; linear regression in section 2.1.1, Poisson regression in section 2.1.2, and Bernoulli regression in section 2.1.3.

We amend the exponential family notation to read

$$\exp\left\{\frac{y_{it}\theta_{it} - b(\theta_{it})}{a(\phi)} - c(y_{it}, \phi)\right\} \tag{2.68}$$

where the repeated observations $t = 1, \ldots, n_i$ within a given panel i are assumed to be correlated. GLMs assume that the observations are independent with no correlation between the outcomes. Marginal models, GEE models, and random-effects models are extensions of the GLM for correlated data. In the next few sections we illustrate many of the methods for addressing the correlation inherent in panel data.

Throughout these subsections, we include results for analyzing the ship (see section 5.2.2) and wheeze (see section 5.2.1) data. We model the ship data using panel Poisson estimators; the wheeze data are modeled using panel (logistic) binomial estimators.

2.3.1 Pooled estimators

A simple approach to modeling panel data is simply to ignore the panel dependence that might be present in the data. The result of this approach is called a *pooled* estimator, since the data are simply pooled without regard to which panel the data naturally belong. The resulting estimated coefficient vector, though consistent, is not efficient. A direct result of ignoring the within-panel correlation is that the estimated (naive) standard errors are not a reliable measure for testing purposes. To address the standard errors, we should employ a modified sandwich estimate of variance, or another variance estimate that adjusts for the panel nature of the data.

The general LIML exponential family pooled estimating equation is given by

$$\left[\left\{ \frac{\partial \mathcal{L}}{\partial \beta_j} = \sum_{i=1}^{n} \sum_{t=1}^{n_i} \frac{y_{it} - \mu_{it}}{a(\phi) \mathrm{V}(\mu_{it})} \left(\frac{\partial \mu}{\partial \eta}\right)_{it} x_{jit} \right\}_{j=1,\ldots,p} \right]_{p \times 1} = [\mathbf{0}]_{p \times 1}$$
(2.69)

where p is the column dimension of the matrix of covariates $\mathbf{X}$. Apart from a second subscript, the pooled estimating equation is no different from equation 2.55. The implied likelihood of the estimating equation does not address any second order dependence of the data. In other words, if we believe that there is within-panel dependence, our implied likelihood is wrong.

Since our estimating equation does not imply a likelihood that includes within-panel dependence of the data, we must be very careful in our interpretation of results. The usual variance matrix $\widehat{\mathrm{V}}_\mathrm{H}$ obtained from fitting the GLM is naive in the sense that it assumes no within-panel dependence of the data. Instead, we can use the modified sandwich estimate of variance for testing and interpretation; but we should acknowledge the fact that employing a pooled estimator with a modified variance estimate is a declaration that the underlying likelihood is not correct.

The modified sandwich estimate of variance addresses possible within-panel correlation, as in Binder (1983), by summing residuals over the panel identifiers in the estimation of the variance of the estimating equation. However, it does not alter the estimating equation itself. Therefore, the implied likelihood of the estimating equation is unchanged; it does not directly address within panel correlation, nor does it change the resulting coefficient estimates from a hypothesis of within panel independence.

Rather, the modified sandwich estimate of variance alters the variance estimate. Our interpretation of coefficients is in terms of an underlying best-fit independence model for data that in fact come from a dependence model. In other words, there is a best independence model for the data consisting of the entire population of panels and observations; our results estimate this best independence model.

In the sense that we have explicitly addressed possible within panel correlation without altering the estimating equation from the independence model, we are fitting a model from the wrong (implied) likelihood. It is for this reason that researchers using this variance estimate do not use likelihood-based criteria and tests to interpret the model.

The modified sandwich estimate of variance is robust to *any* type of correlation within panels. Many people believe that the adjective *robust* means that sandwich estimates of variance are larger than naive estimates of variance. This is not the case. A robust variance estimate may result in smaller or larger estimators depending on the nature of the within-panel correlation. The calculation of the modified sandwich estimate of variance uses the sums of the residuals from each panel. If the residuals are negatively correlated and the sums are small, the modified sandwich estimate of variance produces smaller standard errors than the naive estimator.

Declaring that the underlying likelihood of the fitted model is not correct by using the modified sandwich estimate of variance requires more than a careful interpretation of model results. One must also be vigilant about not employing model fit diagnostics and tests based on likelihood calculations or assumptions. There is no free lunch with this variance adjustment. One should not adjust the variance due to a belief that there is a violation of independence of the observations, and then ignore this fact later; for example, by running a likelihood ratio test comparing a nested model. That is, we can not use these post-estimation tests and diagnostics outside of the interpretation of our model being an estimate of some incorrect best fit independence model of a population of observations.

2.3.2 Fixed-effects and random-effects models

To address the panel structure in our data, we may include an effect for each panel in our estimating equation. We may assume that these effects are fixed effects or random effects. In addition, the fixed effects may be conditional fixed effects or unconditional fixed effects. Unconditional fixed-effects estimators simply include an indicator variable for the panel in our estimation. Conditional fixed-effects estimators are derived from a different likelihood. They are derived from a conditional likelihood, which removes the fixed effects from the estimation by conditioning on the sufficient statistic for the parameter to be removed.

There is some controversy over the choice of fixed-effects or random-effects models. The choice is clear when the nature of the panels is known. The inference follows the nature of the model. When there is no compelling choice between the two models, the random-effects model is sometimes preferred if there are covariates that are constant within panels. Coefficients for these covariates can not be estimated for fixed-effects models since the covariate is collinear with the fixed effect.

2.3.2.1 Unconditional fixed-effects models

If there are a finite number of panels in a population and each panel is represented in our sample, we would use an unconditional fixed-effects model. If there are an infinite number of panels (or effectively uncountable), then we would use a conditional fixed-effects model, because using an unconditional fixed-effects model would result in biased estimates.

The unconditional fixed-effects estimating equation for the exponential family is given by admitting the fixed effect ν_i into the linear predictor $\eta_{it} = \mathbf{x}_{it}\boldsymbol{\beta} + \nu_i$ where $\mathbf{x}_{it}$ is the itth row of the $\mathbf{X}$ matrix. We wish to estimate the $(p+n) \times 1$ parameter vector $\Theta = (\boldsymbol{\beta}, \boldsymbol{\nu})$. The estimating equation for the unconditional fixed-effects GLM is given by

$$\left[\begin{array}{l} \left\{ \dfrac{\partial \mathcal{L}}{\partial \beta_j} = \sum_{i=1}^{n}\sum_{t=1}^{n_i} \dfrac{y_{it}-\mu_{it}}{a(\phi)\mathrm{V}(\mu_{it})} \left(\dfrac{\partial \mu}{\partial \eta}\right)_{it} x_{jit} \right\} \\ \left\{ \dfrac{\partial \mathcal{L}}{\partial \nu_k} = \sum_{t=1}^{n_k} \dfrac{y_{kt}-\mu_{kt}}{a(\phi)\mathrm{V}(\mu_{kt})} \left(\dfrac{\partial \mu}{\partial \eta}\right)_{kt} \right\} \end{array} \right]_{(p+n)\times 1} = [\mathbf{0}]_{(p+n)\times 1}$$

(2.70)

for $j = 1, \ldots, p$ and $k = 1, \ldots, n$.

Unconditional fixed-effects models may be obtained for the full complement of GLMs including those implying quasilikelihoods.

Using the ship data, we fit an unconditional fixed effects Poisson model by including indicator variables for the ship. The results are given by

```
Poisson regression                              Number of obs   =         34
                                                LR chi2(8)      =     107.63
                                                Prob > chi2     =     0.0000
Log likelihood = -68.280771                     Pseudo R2       =     0.4408

------------------------------------------------------------------------------
    incident |      Coef.   Std. Err.      z    P>|z|     [95% Conf. Interval]
-------------+----------------------------------------------------------------
    _Iship_2 |  -.5433443   .1775899    -3.06   0.002    -.8914141   -.1952745
    _Iship_3 |  -.6874016   .3290472    -2.09   0.037    -1.332322   -.042481
    _Iship_4 |  -.0759614   .2905787    -0.26   0.794    -.6454851    .4935623
    _Iship_5 |   .3255795   .2358794     1.38   0.168    -.1367357    .7878946
    op_75_79 |   .384467    .1182722     3.25   0.001     .1526578    .6162761
    co_65_69 |   .6971404   .1496414     4.66   0.000     .4038487    .9904322
    co_70_74 |   .8184266   .1697736     4.82   0.000     .4856763   1.151177
    co_75_79 |   .4534266   .2331705     1.94   0.052    -.0035791    .9104324
       _cons |  -6.405902   .2174441   -29.46   0.000    -6.832084   -5.979719
    exposure |  (offset)
------------------------------------------------------------------------------
```

Using the wheeze data, we fit an unconditional fixed-effects logistic regression model by including indicator variables for the child (case). The results are given by

```
Logit estimates                                   Number of obs   =         40
                                                  LR chi2(11)     =       8.44
                                                  Prob > chi2     =     0.6731
Log likelihood = -23.454028                       Pseudo R2       =     0.1525

------------------------------------------------------------------------------
    wheeze |      Coef.   Std. Err.      z     P>|z|     [95% Conf. Interval]
-----------+------------------------------------------------------------------
  _Icase_3 |  -.9845011   1.931727    -0.51    0.610    -4.770617    2.801615
  _Icase_4 |  -1.167123   1.571454    -0.74    0.458    -4.247117    1.912871
  _Icase_9 |  -2.462266   2.221064    -1.11    0.268    -6.815471    1.890939
 _Icase_11 |   1.321318   2.109374     0.63    0.531    -2.812979    5.455615
 _Icase_13 |   1.253285   1.794704     0.70    0.485    -2.264269     4.77084
 _Icase_14 |  -2.446925   2.137959    -1.14    0.252    -6.637248    1.743397
 _Icase_15 |  -1.073333   1.621803    -0.66    0.508    -4.252007    2.105342
 _Icase_16 |  -1.284241   1.925523    -0.67    0.505    -5.058197    2.489715
  kingston |  -1.341676   2.210786    -0.61    0.544    -5.674736    2.991385
       age |  -.3607877   .3399243    -1.06    0.289    -1.027027    .3054517
     smoke |   .1154117   .8828089     0.13    0.896    -1.614862    1.845685
     _cons |   4.927258   3.828498     1.29    0.198     -2.57646    12.43098
------------------------------------------------------------------------------
```

In estimating the model, it is determined that several of the indicator variables for the subject predict the outcome perfectly. In such a case, software may drop these variables (as the outcome above reflects). Keeping these perfect predictors in the model requires (in maximum likelihood) that the fitted coefficients should be infinite. An alternative would be to use software (or programming techniques) that model exact logistic regression.

2.3.2.2 Conditional fixed-effects models

A conditional fixed-effects model is formed by conditioning out the fixed effects from the estimation. This allows a much more efficient estimator at the cost of placing constraints on inference in the form of the conditioning imposed on the likelihood. Such models are derived from specific distributions with valid likelihoods.

Conditional fixed-effects models are derived from specific distributions, not from the general exponential family distribution. For illustration of the model construction, we derive the estimating equation for the FIML conditional fixed effects Poisson regression model. Apart from identifying a sufficient statistic on which to condition, the derivation of the estimating equation is the same as for the previous illustrations for independent data.

In general, we have a specific distribution for a single outcome on y_{it} that we call $f_1(y_{it})$. We find the joint distribution for all of the observations for a specific panel $f_1(\mathbf{y}_i) = \prod_{t=1}^{n_i} f_1(y_{it})$ and obtain the sufficient statistic $\xi(\mathbf{y}_i)$ for the fixed effect ν_i. We then find the distribution of the sufficient statistic $f_2(\xi(\mathbf{y}_i))$. Finally, we obtain the conditional distribution of the outcomes given the distribution of the sufficient statistic as $f_3(\mathbf{y}_i;\boldsymbol{\beta}|\xi(\mathbf{y}_i)) = f_1(\mathbf{y}_i)/f_2(\xi(\mathbf{y}_i))$. This distribution is free of the fixed effect ν_i. Thus, the conditional log-likelihood

PANEL DATA

for all of the panels is given by

$$\mathcal{L} = \ln \prod_{i=1}^{n} f_3(\mathbf{y}_i; \boldsymbol{\beta} | \xi(\mathbf{y}_i)) \qquad (2.71)$$

with estimating equation for $\Theta = (\boldsymbol{\beta})$ and $j = 1, \ldots, p$

$$\Phi(\Theta) = \left[\left\{ \frac{\partial \mathcal{L}}{\partial \beta_j} \right\} \right]_{(p \times 1)} = [\mathbf{0}]_{(p \times 1)} \qquad (2.72)$$

The estimating equation can be made FIML or LIML depending on whether there are additional parameters from f_1 that we include in Θ or treat as ancillary. We shall derive the conditional fixed-effects Poisson regression model to highlight the steps outlined above. Our specific model will assume the canonical log link for the relationship of the linear predictor to the expected value.

The probability for a specific outcome in the (individual level) Poisson model is

$$P(Y_{it} = y_{it}) = e^{-\mu_{it}} \mu_{it}^{y_{it}} / y_{it}! \qquad (2.73)$$

where the expected value of the outcome is given by μ_{it}. As done previously, we specify the relationship of the expected value to the linear predictor through a link function; in this case, the log link.

$$\mu_{it} = \exp(x_{it}\boldsymbol{\beta} + \gamma_i) = \exp(\eta_{it} + \gamma_i) \qquad (2.74)$$

Here, we have made the inclusion of the fixed effect γ_i explicit in the parameterization. The probability for a specific outcome, introducing covariates and the fixed effect, is

$$P(Y_{it} = y_{it}) = e^{-\exp(\eta_{it} + \gamma_i)} \exp(\eta_{it} + \gamma_i)^{y_{it}} / y_{it}! \qquad (2.75)$$

Since observations within a panel are independent, we write the probability of a vector of outcomes for panel i as

$$P(\mathbf{Y}_i = \mathbf{y}_i) = \prod_{t=1}^{n_i} e^{-\exp(\eta_{it} + \gamma_i)} \exp(\eta_{it} + \gamma_i)^{y_{it}} / y_{it}! \qquad (2.76)$$

$$= e^{-\sum_t \exp(\eta_{it} + \gamma_i)} \exp(\gamma_i)^{\sum_t y_{it}} \prod_{t=1}^{n_i} \frac{\exp(\eta_{it} y_{it})}{y_{it}!} \qquad (2.77)$$

The sufficient statistic for γ_i is then $\xi(\mathbf{y}_i) = \sum_{t=1}^{n_i} y_{it}$.

Since we know that the sum of Poisson random variables is also Poisson, we write the probability of a given outcome for the sufficient statistic as

$$P\left(\sum_{t=1}^{n_i} Y_{it} = \sum_{t=1}^{n_i} y_{it}\right)$$

$$= e^{-\sum_t \exp(\eta_{it}+\gamma_i)} \left(\sum_{t=1}^{n_i} \exp(\eta_{it}+\gamma_i)\right)^{\sum_t y_{it}} \Big/ \left(\sum_{t=1}^{n_i} y_{it}\right)! \quad (2.78)$$

$$= e^{-\sum_t \exp(\eta_{it}+\gamma_i)} \exp(\gamma_i)^{\sum_t y_{it}} \left(\sum_{t=1}^{n_i} \exp(\eta_{it})\right)^{\sum_t y_{it}} \Big/ \left(\sum_{t=1}^{n_i} y_{it}\right)! \quad (2.79)$$

The conditional probability is then the ratio of the joint probability (equation 2.77) and the probability of the sufficient statistic (equation 2.79) given by

$$P\left(\mathbf{Y}_i = \mathbf{y}_i \Big| \sum_{t=1}^{n_i} y_{it}\right) = \frac{(\sum_t y_{it})!}{(\sum_t \exp(\eta_{it}))^{\sum_t y_{it}}} \prod_{t=1}^{n_i} \left(\frac{\exp(\eta_{it} y_{it})}{y_{it}!}\right) \quad (2.80)$$

The conditional probability of observations for a single panel is free of the panel-level fixed effect since we conditioned on the sufficient statistic for the panel-level effect. The estimating equation is derived using the standard approach illustrated throughout this chapter. The conditional likelihood is thus given by the product of the conditional probabilities for all of the panels

$$L(\beta) = \prod_{i=1}^n \frac{(\sum_t y_{it})!}{(\sum_t \exp(\eta_{it}))^{\sum_t y_{it}}} \prod_{t=1}^{n_i} \left(\frac{\exp(\eta_{it} y_{it})}{y_{it}!}\right) \quad (2.81)$$

$$= \exp\left\{\sum_{i=1}^n \ln\Gamma\left(\sum_{t=1}^{n_i} y_{it} + 1\right) - \sum_{t=1}^{n_i} y_{it} \ln(\exp(\eta_{it})) + \sum_{t=1}^{n_i} (\eta_{it} y_{it} - \ln\Gamma(y_{it}+1))\right\} \quad (2.82)$$

Consequently, the log-likelihood is

$$\mathcal{L}(\beta) = \sum_{i=1}^n \ln\Gamma\left(\sum_{t=1}^{n_i} y_{it} + 1\right) - \sum_{t=1}^{n_i} y_{it} \ln(\exp(\eta_{it})) + \sum_{t=1}^{n_i} (\eta_{it} y_{it} - \ln\Gamma(y_{it}+1)) \quad (2.83)$$

For this particular case, there are no ancillary parameters and the estimating equation $\Psi(\beta) = \mathbf{0}$ for the conditional fixed effects log-linked Poisson model is the derivative of the above log-likelihood for $j = 1,\ldots,p$

$$\left[\left\{\frac{\partial \mathcal{L}}{\partial \eta}\frac{\partial \eta}{\partial \beta_j} = \sum_{i=1}^n \sum_{t=1}^{n_i} \left[y_{it} - y_{it} \frac{\exp(\eta_{it})}{\sum_k \exp(\eta_{ik})}\right] x_{jit}\right\}\right]_{(p\times 1)} = [\mathbf{0}]_{(p\times 1)} \quad (2.84)$$

Fitting a conditional fixed effects model using the ship data provides the following results:

```
Conditional fixed-effects Poisson          Number of obs       =        34
Group variable (i) : ship                  Number of groups    =         5

                                           Obs per group: min =         6
                                                          avg =       6.8
                                                          max =         7

                                           Wald chi2(4)        =     48.44
Log likelihood  = -54.641859               Prob > chi2         =    0.0000

------------------------------------------------------------------------------
    incident |      Coef.   Std. Err.      z    P>|z|     [95% Conf. Interval]
-------------+----------------------------------------------------------------
    op_75_79 |   .384467    .1182722     3.25   0.001     .1526578    .6162761
    co_65_69 |   .6971405   .1496414     4.66   0.000     .4038487    .9904322
    co_70_74 |   .8184266   .1697737     4.82   0.000     .4856764    1.151177
    co_75_79 |   .4534267   .2331705     1.94   0.052    -.0035791    .9104324
    exposure |  (offset)
------------------------------------------------------------------------------
```

The same logic can be used to derive the conditional fixed-effects logistic regression model. First, the probability of a given outcome is specified as

$$P(Y_{it} = y_{it}) = \mu_{it}^{y_{it}} (1 - \mu_{it})^{1-y_{it}} \tag{2.85}$$

Introducing the covariates and the fixed effect through the canonical logit link function yields

$$\mu_{it} = \frac{\exp(x_{it}\beta + \gamma_i)}{1 + \exp(x_{it}\beta + \gamma_i)} = \frac{\exp(\eta_{it} + \gamma_i)}{1 + \exp(\eta_{it} + \gamma_i)} \tag{2.86}$$

The probability of a given outcome is

$$P(Y_{it} = y_{it}) = \left(\frac{\exp(\eta_{it} + \gamma_i)}{1 + \exp(\eta_{it} + \gamma_i)}\right)^{y_{it}} \left(\frac{1}{1 + \exp(\eta_{it} + \gamma_i)}\right)^{1-y_{it}} \tag{2.87}$$

$$= \exp\{y_{it}(\eta_{it} + \gamma_i) - \ln(1 + \exp(\eta_{it} + \gamma_i))\} \tag{2.88}$$

Since observations within a panel are independent, the probability of a vector of outcomes for panel i is the product

$$P(\mathbf{Y}_i = \mathbf{y}_i)$$

$$= \prod_{t=1}^{n_i} \exp\{y_{it}(\eta_{it} + \gamma_i) - \ln(1 + \exp(\eta_{it} + \gamma_i))\} \tag{2.89}$$

$$= \exp\left\{\sum_{t=1}^{n_i} y_{it}\eta_{it} + \gamma_i \sum_{t=1}^{n_i} y_{it} - \sum_{t=1}^{n_i} \ln(1 + \exp(\eta_{it} + \gamma_i))\right\} \tag{2.90}$$

The sufficient statistic for γ_i is then $\xi(\mathbf{y}_i) = \sum_{t=1}^{n_i} y_{it}$.

We know the ratio of the joint distribution and the distribution of the sufficient statistic $\xi(\mathbf{y}_i)$ does not involve the fixed effect γ_i. Unfortunately, the sum of Bernoulli random variables, when the individual observations do not all have the same probability of success, is not characterized by a known distribution (as in the previous example for the conditional fixed effects Poisson model).

Imagine that we have $X_1 \sim$ Bernoulli(p_1), $X_2 \sim$ Bernoulli(p_2), $X_3 \sim$ Bernoulli(p_3), and we desire the probability distribution of $T = X_1 + X_2 + X_3$. The possible outcomes and associated probabilities for the sum of these Bernoulli random variables is given by

$$P(T = 0) = (1 - p_1)(1 - p_2)(1 - p_3) \qquad (2.91)$$
$$P(T = 1) = (p_1)(1 - p_2)(1 - p_3)$$
$$+ (1 - p_1)(p_2)(1 - p_3)$$
$$+ (1 - p_1)(1 - p_2)(p_3) \qquad (2.92)$$
$$P(T = 2) = (p_1)(p_2)(1 - p_3)$$
$$+ (p_1)(1 - p_2)(p_3)$$
$$+ (1 - p_1)(p_2)(p_3) \qquad (2.93)$$
$$P(T = 3) = (p_1)(p_2)(p_3) \qquad (2.94)$$

Since this probability distribution is not a simple known distribution for which we can easily look up a single formula, we must derive a useful and workable characterization. For this rather elementary example of a sum of three Bernoulli random variables, we imagine a vector of indicator variables $\mathbf{d} = (d_1, d_2, d_3)$ where $d_i \in \{0, 1\}$ specifies whether the random variable X_i is a success ($d_i = 1$) or failure ($d_i = 0$). For outcome $T = k$, we let S_k denote the set of vectors d such that $\sum_i d_i = k$. Clearly, the number of terms in S_k is given by $\binom{3}{k}$ for outcome equal to k.

T	S_k
0	$\{(0,0,0)\}$
1	$\{(1,0,0), (0,1,0), (0,0,1)\}$
2	$\{(1,1,0), (1,0,1), (0,1,1)\}$
3	$\{(1,1,1)\}$

Using this notation, we can then construct the conditional probabilities as

$$P(X_1 = x_1, X_2 = x_2, X_3 = x_3 | T = k)$$
$$= \frac{p_1^{x_1}(1-p_1)^{1-x_1} p_2^{x_2}(1-p_2)^{1-x_2} p_3^{x_3}(1-p_3)^{1-x_3}}{\sum_{\mathbf{d} \in S_k} \prod_{i=1}^{3} p_i^{d_i}(1-p_i)^{1-d_i}} \qquad (2.95)$$

For example,

$$P(X_1 = 1, X_2 = 0, X_3 = 0 | T = 1)$$
$$= \frac{p_1(1-p_2)(1-p_3)}{p_1(1-p_2)(1-p_3) + (1-p_1)p_2(1-p_3) + (1-p_1)(1-p_2)p_3} \qquad (2.96)$$

PANEL DATA

In our more general case, we add a subscript i to the above notation to reflect the association to panel i. For a given conditional outcome $\sum_t y_{it} = k_i$, there are $\binom{n_i}{k_i}$ possible terms (generalizing from the simple example above). Let $\mathbf{d}_i$ denote a vector of indicators of length n_i indicating whether a particular observation is a success. Let S_{k_i} denote the set of vectors $\mathbf{d}_i$ such that $\sum_t d_{it} = k_i$.

The conditional probability (from our logit-link parameterization) is then given by

$$\frac{\exp(y_{it}\eta_{it})}{\sum_{\mathbf{d}_i \in S_{k_i}} \exp(d_{it}\eta_{it})} \tag{2.97}$$

The conditional likelihood is given by the product of the conditional probabilities for all of the panels

$$L(\boldsymbol{\beta}) = \prod_{i=1}^{n} \frac{\exp(y_{it}\eta_{it})}{\sum_{\mathbf{d}_i \in S_{k_i}} \exp(d_{it}\eta_{it})} \tag{2.98}$$

$$= \sum_{i=1}^{n} \exp\left\{ y_{it}\eta_{it} - \ln \sum_{\mathbf{d}_i \in S_{k_i}} \exp(d_{it}\eta_{it}) \right\} \tag{2.99}$$

Finally, the conditional log-likelihood is

$$\mathcal{L}(\boldsymbol{\beta}) = \sum_{i=1}^{n} \left\{ y_{it}\eta_{it} - \ln \sum_{\mathbf{d}_i \in S_{k_i}} \exp(d_{it}\eta_{it}) \right\} \tag{2.100}$$

from which the estimating equation can be derived as the derivative of the log-likelihood for $j = 1, \ldots, p$

$$\left[\left\{ \Psi(\boldsymbol{\beta}) = \frac{\partial}{\partial \eta} \frac{\partial \eta}{\partial \beta_j} \sum_{i=1}^{n} \left(\sum_{t=1}^{n_i} y_{it}\eta_{it} - \ln \sum_{\mathbf{d}_i \in S_{k_i}} \exp(d_{it}\eta_{it}) \right) \right\} \right]_{(p \times 1)} = [\mathbf{0}]_{(p \times 1)} \tag{2.101}$$

We emphasize that the conditional probability in equation 2.97 is really the ratio of a fraction to the sum of fractions. Each fractional term has a common denominator that cancels and was suppressed in printing the equation.

Using the binary outcome wheeze data, we estimate a conditional fixed-effects logistic regression model. The results are given by

```
Conditional fixed-effects logit          Number of obs     =         40
Group variable (i) : case                Number of groups  =         10

                                         Obs per group: min =          4
                                                        avg =        4.0
                                                        max =          4

                                         LR chi2(2)        =       0.91
Log likelihood = -14.622988              Prob > chi2       =     0.6336
```

```
------------------------------------------------------------------------
 wheeze |      Coef.   Std. Err.      z    P>|z|     [95% Conf. Interval]
--------+---------------------------------------------------------------
    age | -.2701682    .2938288    -0.92   0.358    -.8460621    .3057256
  smoke |  .0900261    .7720841     0.12   0.907    -1.423231    1.603283
------------------------------------------------------------------------
```

The results indicate that only 40 observations are used instead of the full 64 observations. In these data, there are 6 cases for which the outcome does not vary for the child; the sum of successful outcomes is either zero or n_i so that the denominator in equation 2.97 has only one term. As such, the conditional probability for the observations, given the sum of the outcomes, is equal to one. Since the log of this outcome is zero ($\ln 1 = 0$), there is no contribution to the log-likelihood calculation, and the subjects are dropped from the estimation. In reality, they could remain in the estimation; but since they contribute no information to the model estimation, there is no reason to artificially increase the sample size. For illustration, in our simple example of the sum of three Bernoulli random variables, note that the conditional probabilities for $T = 0$ and $T = 3$ are both equal to one.

Having panels with a conditional probability equal to one can also occur in Poisson models. A Poisson model for which all of the outcomes in a panel are zero conditions on the sum of the outcomes being zero. Since there is only one possible set of outcomes for the individual measurements in the panel, the conditional probability of the panel is one. In other words, if the sum of the outcomes in a panel of size n_i is equal to zero, the conditional probability

$$P\left(y_{i1} = y_{i2} = \ldots = y_{in_i} = 0 \;\middle|\; \sum_{t=1}^{n_i} y_{it} = 0\right) = 1 \qquad (2.102)$$

In such cases, we recommend dropping those panels from the conditional fixed-effects Poisson model just as those panels were dropped in the conditional fixed-effects logistic model above. We emphasize that there were no such panels in the ship data illustrated earlier.

2.3.2.3 Random-effects models

A random-effects model parameterizes the random effects according to an assumed distribution for which the parameters of the distribution are estimated. These models are called subject-specific models, since the likelihood models the individual observations instead of the marginal distribution of the panels. As in the case of conditional fixed-effects models, our derivation begins with an assumed distribution and, thus, does not address the quasilikelihoods of GLMs.

The log-likelihood for a random-effects model is

$$\mathcal{L} = \ln \prod_{i=1}^{n} \int_{-\infty}^{\infty} f(\nu_i) \left\{ \prod_{t=1}^{n_i} f_y(\mathbf{x}_{it}\beta + \nu_i) \right\} d\nu_i \qquad (2.103)$$

PANEL DATA

where f_y is the assumed density for the overall model (the outcome) and f is the density of the iid random effects ν_i. The estimating equation is the derivative of the log-likelihood in terms of β and the parameters of the assumed random-effects distribution.

By inspection, obtaining the estimating equation might be a formidable task. There are cases for which an analytic solution of the integral is possible and for which the resulting estimating equation may be easily calculated. This depends on both the distribution of the outcome variable and the distribution of the random effect. There are also cases for which numeric integration techniques, e.g., quadrature formulae, may be implemented in order to calculate the estimating equation. In the following, we present an example of each of these approaches.

Revisiting the Poisson setting, a random effects model may be derived assuming a gamma distribution for the random effect. This choice of distribution leads to an analytic solution of the integral in the likelihood.

In the usual Poisson model we hypothesize that the mean of the outcome variable y is given by $\lambda_{it} = \exp(\mathbf{x}_{it}\beta)$. In the panel setting we assume that each panel has a different mean that is given by $\exp(\mathbf{x}_{it}\beta + \eta_i) = \lambda_{it}\nu_i$. As such, we refer to the random effect as entering multiplicatively rather than additively, as is the case in random-effects linear regression.

Since the random effect $\nu_i = \exp(\eta_i)$ is positive, we select a gamma distribution adding the restriction that the mean of the random effects equals one. We do this so that there is only one additional parameter θ to estimate.

$$f(\nu_i) = \frac{\theta^\theta}{\Gamma(\theta)} \nu_i^{\theta-1} \exp(-\theta \nu_i) \qquad (2.104)$$

The conditional mean of the outcome given the random effect is Poisson, and the random effect is distributed Gamma(θ, θ). Therefore, we take the product to obtain the joint density function for the observations of a single panel given by

$$f(\nu_i, \lambda_{i1}, \ldots, \lambda_{in_i}) = \frac{\theta^\theta}{\Gamma(\theta)} \nu_i^{\theta-1} \exp(-\theta \nu_i) \prod_{t=1}^{n_i} \exp(-\nu_i \lambda_{it})(\nu_i \lambda_{it})^{y_{it}}/y_{it}! \qquad (2.105)$$

Moreover, since the panels are all independent, the joint density for all of the panels combined is the product of the density of each of the panels.

The log-likelihood for gamma distributed random effects may then be derived by integrating over ν_i. We note that by rearranging terms in the joint density, the integral term may be simplified to one since it is the integral of another gamma random variable. After simplification and collection of terms, we substitute our preferred μ_i notation for the expected value λ for consistency and to address the goal of introducing covariates. The log-likelihood is then specified as

$$\mathcal{L} = \sum_{i=1}^{n} \left\{ \ln \Gamma \left(\theta + \sum_{t=1}^{n_i} y_{it} \right) - \ln \Gamma(\theta) - \sum_{t=1}^{n_i} \ln \Gamma(y_{it} + 1) + \theta \ln u_i \right.$$

$$+ \left(\sum_{t=1}^{n_i} y_{it} \right) \ln(1 - u_i) - \left(\sum_{t=1}^{n_i} y_{it} \right) \ln \left(\sum_{t=1}^{n_i} \mu_{it} \right)$$

$$\left. + \sum_{t=1}^{n_i} y_{it} \ln(\mu_{it}) \right\} \tag{2.106}$$

$$u_i = \frac{\theta}{\theta + \sum_{t=1}^{n_i} \mu_{it}} \tag{2.107}$$

and $\mu_{it} = \exp(\mathbf{x}_{it}\boldsymbol{\beta})$.

The estimating equation $\Psi(\Theta) = \Psi(\boldsymbol{\beta}, \theta)$ for a gamma distributed random effects Poisson model is then given by setting the derivative of the log-likelihood to zero

$$\begin{bmatrix} \left\{ \frac{\partial \mathcal{L}}{\partial \beta_j} \right\} \\ \left\{ \frac{\partial \mathcal{L}}{\partial \theta} \right\} \end{bmatrix}_{(p+1) \times 1} = [\mathbf{0}]_{(p+1) \times 1} \tag{2.108}$$

where

$$\frac{\partial \mathcal{L}}{\partial \beta_j} = \sum_{i=1}^{n} \sum_{t=1}^{n_i} x_{jit} \left[y_{it} + \mu_{it} \left((u_i - 1) \frac{\sum_{\ell=1}^{n_i} y_{i\ell}}{\sum_{\ell=1}^{n_i} \mu_{i\ell}} - u_i \right) \right] \left(\frac{\partial \mu}{\partial \eta} \right)_{it} \tag{2.109}$$

$$\frac{\partial \mathcal{L}}{\partial \theta} = \sum_{i=1}^{n} \left[\psi \left(\theta + \sum_{t=1}^{n_i} y_{it} \right) - \psi(\theta) + \ln u_i + (1 - u_i) - \frac{u_i}{\theta} \sum_{t=1}^{n_i} y_{it} \right] \tag{2.110}$$

and u_i is defined in equation 2.107. In the derivative with respect to θ (equation 2.110), we use $\psi()$ to denote the derivative of the log of the Gamma function (the psi-function). This is a standard notation for this function and should not be confused with our use of $\Psi()$ (capital psi) in other sections to denote the estimating equation.

Using the ship data, we fit a gamma distributed random effects Poisson model. In this case, there is no need to approximate the likelihood through quadrature (or any other means). Instead, there is an analytic solution to the likelihood despite the need to integrate over the random effect. This is the real benefit of choosing the gamma distribution for the random effect in a Poisson model.

The results of fitting a gamma distributed random-effects model for the ship data are presented as

PANEL DATA

```
Random-effects Poisson                          Number of obs      =         34
Group variable (i) : ship                       Number of groups   =          5

Random effects u_i ~ Gamma                      Obs per group: min =          6
                                                               avg =        6.8
                                                               max =          7

                                                Wald chi2(4)       =      50.90
Log likelihood  = -74.811217                    Prob > chi2        =     0.0000

------------------------------------------------------------------------------
   incident |      Coef.   Std. Err.       z    P>|z|     [95% Conf. Interval]
------------+-----------------------------------------------------------------
   op_75_79 |   .3827453   .1182568     3.24   0.001      .1509662    .6145244
   co_65_69 |   .7092879   .1496072     4.74   0.000      .4160633    1.002513
   co_70_74 |   .8573273   .1696864     5.05   0.000      .5247481    1.189906
   co_75_79 |   .4958618   .2321316     2.14   0.033      .0408922    .9508313
      _cons |  -6.591175   .2179892   -30.24   0.000     -7.018426   -6.163924
   exposure |   (offset)
------------+-----------------------------------------------------------------
   /lnalpha |  -2.368406   .8474597                      -4.029397   -.7074155
------------+-----------------------------------------------------------------
      alpha |   .0936298   .0793475                       .0177851    .4929165
------------------------------------------------------------------------------
Likelihood ratio test of alpha=0: chibar2(01) =   10.61 Prob>=chibar2 = 0.001
```

Applying the well-known Gauss–Hermite quadrature approximation, a common random-effects model can be derived for Gaussian distributed random effects. The likelihood is based on the joint distribution of the outcome and the Gaussian random effect. After completing the square of terms in the model, the resulting likelihood is the product of functions of the form

$$\int_{-\infty}^{\infty} e^{-z^2} f(z) dz \qquad (2.111)$$

This may be numerically approximated using the Gauss–Hermite quadrature formula. The accuracy of the approximation is affected by the number of points used in the quadrature calculation and the smoothness of the product of the functions $f(z_i)$—how well this product may be approximated by a polynomial.

Applying this approach to the construction of a Gaussian random-effects Poisson regression model, we obtain a quadrature approximated log-likelihood $\mathcal{L}_a$ formulated as

$$\mathcal{L}_a = \sum_{i=1}^{n} \ln \frac{1}{\sqrt{\pi}} \sum_{m=1}^{M} w_m^* \prod_{t=1}^{n_i} \mathcal{F}\left(\mathbf{x}_{it}\beta + \sqrt{2\frac{\rho}{1-\rho}}\, x_m^*\right) \qquad (2.112)$$

where (w_m^*, x_m^*) are the quadrature weights and abscissa, M is the number of points used in the quadrature rule, and $\rho = \sigma_\nu^2/(\sigma_\nu^2 + \sigma_\epsilon^2)$ is the proportion of total variance contributed by the random effect variance component. For the Poisson model of interest,

$$\mathcal{F}(z) = \exp\{-\exp(z)\} \exp(z)^{y_{it}} / y_{it}! \qquad (2.113)$$

The estimating equation for this likelihood-based model is specified by setting the derivative of the log-likelihood to zero.

Using the ship data, we fit a Gaussian distributed random effects Poisson model. The results are given by

```
Random-effects poisson                          Number of obs      =        34
Group variable (i) : ship                       Number of groups   =         5

Random effects u_i ~ Gaussian                   Obs per group: min =         6
                                                               avg =       6.8
                                                               max =         7

                                                LR chi2(4)         =     55.93
Log likelihood    = -74.225924                  Prob > chi2        =    0.0000

------------------------------------------------------------------------------
    incident |      Coef.   Std. Err.      z    P>|z|     [95% Conf. Interval]
-------------+----------------------------------------------------------------
     op_75_79|   .3853861   .1182126     3.26   0.001     .1536936    .6170786
     co_65_69|   .7059975   .1495677     4.72   0.000     .4128502    .9991449
     co_70_74|   .8486468   .1695192     5.01   0.000     .5163953    1.180898
     co_75_79|   .4950771   .2302197     2.15   0.032     .0438548    .9462994
       _cons |  -6.732638   .1404479   -47.94   0.000    -7.007911   -6.457365
     exposure|  (offset)
-------------+----------------------------------------------------------------
    /lnsig2u |  -1.42662    .5613872    -2.54   0.011    -2.526919   -.3263217
-------------+----------------------------------------------------------------
     sigma_u |   .4900195   .1375453                      .2826744    .8494545
         rho |   .1936258   .0876521                      .0739925    .4191359
------------------------------------------------------------------------------
Likelihood ratio test of rho=0:   chibar2(01) =    11.78 Prob>=chibar2 = 0.000
```

Using the wheeze data, we fit a Gaussian distributed random effects logistic regression model. The results are given by

```
Random-effects logit                            Number of obs      =        64
Group variable (i) : case                       Number of groups   =        16

Random effects u_i ~ Gaussian                   Obs per group: min =         4
                                                               avg =       4.0
                                                               max =         4

                                                Wald chi2(3)       =      0.93
Log likelihood    = -37.20499                   Prob > chi2        =    0.8170

------------------------------------------------------------------------------
      wheeze |      Coef.   Std. Err.      z    P>|z|     [95% Conf. Interval]
-------------+----------------------------------------------------------------
    kingston |   .1652582   .8476326     0.19   0.845    -1.496071    1.826588
         age |  -.2540051    .282497    -0.90   0.369     -.807689    .2996789
       smoke |  -.0699977   .5360669    -0.13   0.896     -1.12067    .9806742
       _cons |   1.541053   2.931209     0.53   0.599    -4.204012    7.286118
-------------+----------------------------------------------------------------
    /lnsig2u |   .2538943   1.041892                     -1.788176    2.295964
-------------+----------------------------------------------------------------
     sigma_u |   1.135357   .5914594                      .4089805    3.151826
         rho |   .2815162   .0640567                      .0483826    .7512176
------------------------------------------------------------------------------
Likelihood ratio test of rho=0:   chibar2(01) =     2.53 Prob >= chibar2 = 0.056
```

It is worth emphasizing that the random effects are not estimated in these models. Rather, the parameters (variance components) of the assumed distribution of the random effects enter the model. While the approach outlined for Gaussian random effects allows a general specification, one should use caution when assessing models fitted by straight Gauss–Hermite quadrature. The abscissa in this approach are spaced about zero, which may be a poor choice of value for the function to be approximated.

The ease with which one may program Gaussian random effects models has made estimators readily available in software. However, we caution that the Gauss–Hermite quadrature approach does not always provide a good approximation. Better approximations come from *adaptive* quadrature methods that choose abscissas based on the function to be evaluated. At the very least, you should compare results from Gauss–Hermite quadrature approximated models for various numbers of quadrature points to evaluate the stability of the results. Adaptive quadrature approaches can be much better for these types of random-effects models as investigated in Rabe-Hesketh, Skrondal, and Pickles (2002).

Using an adaptive quadrature optimization routine to fit the Gaussian distributed random effects logistic regression model for the wheeze data results in

```
log likelihood = -37.204764
```

wheeze	Coef.	Std. Err.	z	P>|z|	[95% Conf. Interval]
kingston	.1655219	.8478793	0.20	0.845	-1.496291	1.827335
age	-.2540822	.2825297	-0.90	0.368	-.8078303	.2996659
smoke	-.070198	.5360616	-0.13	0.896	-1.120859	.9804634
_cons	1.541421	2.931651	0.53	0.599	-4.204509	7.287352

```
Variances and covariances of random effects
------------------------------------------------------------------------
***level 2 (case)

    var(1): 1.292345 (1.3502367)
------------------------------------------------------------------------
```

where the difference from the straightforward Gauss–Hermite quadrature optimization is apparent. In this particular case, the interpretation of the results does not change and the difference in the fitted coefficients and variance components is not too dramatic. This is not always the case since an adaptive quadrature method can show significant improvement in accuracy. See Rabe-Hesketh et al. (2002) for more information on adaptive quadrature techniques and comparison to nonadaptive optimization.

The difference in results when using adaptive quadrature is more pronounced if we fit a random-effects Poisson model for the Progabide data.

First, let us fit a Gaussian distributed random-effects Poisson model using straightforward Gauss–Hermite quadrature.

```
number of level 1 units = 295
number of level 2 units = 59

log likelihood = -1017.954
```

seizures	Coef.	Std. Err.	z	P>|z|	[95% Conf. Interval]
time	.1118361	.0468766	2.39	0.017	.0199597 .2037125
progabide	.0051622	.0530336	0.10	0.922	-.0987817 .1091062
timeXprog	-.104726	.0650299	-1.61	0.107	-.2321823 .0227303
_cons	1.069857	.0480689	22.26	0.000	.9756434 1.16407
lnPeriod	(offset)				

Variances and covariances of random effects

***level 2 (id)

var(1): .2970534 (.01543218)

Now, let us fit the same model using an adaptive quadrature routine for the estimation.

```
number of level 1 units = 295
number of level 2 units = 59

log likelihood = -1011.0208
```

seizures	Coef.	Std. Err.	z	P>|z|	[95% Conf. Interval]
time	.111836	.0468768	2.39	0.017	.0199591 .203713
progabide	-.0214708	.2101376	-0.10	0.919	-.4333329 .3903914
timeXprog	-.1047258	.0650304	-1.61	0.107	-.232183 .0227315
_cons	1.032649	.1524222	6.77	0.000	.7339074 1.331392
lnPeriod	(offset)				

Variances and covariances of random effects

***level 2 (id)

var(1): .60702391 (.11621224)

Note the increase in the log-likelihood, the change in sign for the progabide coefficient, and the difference in the estimate for the variance of the random effects. In general, the adaptive quadrature results are more accurate than the nonadaptive quadrature results. One can examine the stability of the results for the nonadaptive techniques by fitting the model several times where each estimation uses a different number of quadrature points. If the results are stable, then we can be comfortable with inference for the model.

At the beginning of this section we presented the derivation of random-effects models from an assumed distribution. It is possible to derive a model for random-effects GLM. Estimation may be performed using various optimization techniques including Monte Carlo methods. Zeger and Karim (1991) present a Gibb's sampling approach for constructing GLM random effects models. Basically, the authors describe an estimating equation given by

$$\left[\begin{array}{l} \left\{ \dfrac{\partial \mathcal{L}}{\partial \beta_j} = \sum_{i=1}^{n} \sum_{t=1}^{n_i} \dfrac{y_{it} - \mu_{it}}{a(\phi) V(\mu_{it})} \left(\dfrac{\partial \mu}{\partial \eta} \right)_{it} x_{jit} \right\} \\ \left\{ \dfrac{\partial \mathcal{L}}{\partial \nu_k} = \sum_{t=1}^{n_k} \dfrac{y_{kt} - \mu_{kt}}{a(\phi) V(\mu_{kt})} \left(\dfrac{\partial \mu}{\partial \eta} \right)_{kt} \right\} \end{array} \right]_{(p+n) \times 1} = [\mathbf{0}]_{(p+q) \times 1}$$

(2.114)

for $j = 1, \ldots, p$ and $k = 1, \ldots, q$. The random effects γ_i are assumed to follow some distribution $\mathcal{G}$ characterized by a $(q \times 1)$ parameter vector ν. The authors show that, through the use of conditional distributions, a Monte Carlo approach using Gibb's sampling may be used to estimate the unknown random effects γ_i which are then used to estimate the parameters ν of the distribution of the random effects $\mathcal{G}$. Monte Carlo methods form another class of techniques for constructing and estimating models for panel data.

2.3.3 Population-averaged and subject-specific models

There are two classifications of models that we discuss for addressing the panel structure of data. A population-averaged model is one which includes the within-panel dependence by averaging effects over all panels. A subject-specific model is one which addresses the within-panel dependence by introducing specific panel-level random components.

A population-averaged model, also known as a marginal model, is obtained through introducing a parameterization for a panel-level covariance. The panel-level covariance (or correlation) is then estimated by averaging across information from all of the panels. A subject-specific model is obtained through the introduction of a panel effect. While this implies a panel-level covariance, each panel effect is estimated using information only from the specific panel. Fixed-effects and random-effects models are subject specific.

In the following chapter we further discuss these two classifications and show derivations for subject-specific and population averaged GEE models. These are not the only types of panel data models that one might apply to data. Transitional models and response conditional models are used when the analysis of longitudinal studies must address the dependence of the current response on previous responses. This text does not discuss these models. Interested readers should refer to Neuhaus (1992) for a clear exposition and a useful list of references.

2.4 Estimation

The solution of an estimating equation is obtained using optimization techniques. These techniques iterate toward a solution by updating a current estimate to a new estimate. The common approach employs a Taylor series expansion of an estimating equation given by $\Psi(\beta) = 0$, such that

$$0 = \Psi\left(\beta^{(0)}\right) + \left(\beta - \beta^{(0)}\right)\Psi'\left(\beta^{(0)}\right) + \frac{1}{2}\left(\beta - \beta^{(0)}\right)^2 \Psi''\left(\beta^{(0)}\right) + \cdots \tag{2.115}$$

Keeping only the first two terms, we have the linear approximation

$$0 \approx \Psi\left(\beta^{(0)}\right) + \left(\beta - \beta^{(0)}\right)\Psi'\left(\beta^{(0)}\right) \tag{2.116}$$

$$\beta \approx \beta^{(0)} - \frac{\Psi\left(\beta^{(0)}\right)}{\Psi'\left(\beta^{(0)}\right)} \tag{2.117}$$

Writing this relationship in matrix notation, we then iterate to a solution using the relationship

$$\beta^{(k)} = \beta^{(k-1)} + \left[-\frac{\partial}{\partial \beta}\Psi\left(\beta^{(k-1)}\right)\right]^{-1} \Psi\left(\beta^{(k-1)}\right) \tag{2.118}$$

Thus, given a starting estimate $\beta^{(0)}$, we update our estimate using the relationship in equation 2.118. Specific optimization techniques can take advantage of properties of specific sources of estimating equations. For example, the IRLS algorithm takes advantage of the form of the updating step by using the expected derivative of the estimating equation so that the updating step may be obtained using weighted OLS.

The parameters are estimated separately when there are ancillary parameters in the estimating equation to be solved. This estimation must also be updated at each step. If we consider a second estimating equation for the ancillary parameters, our overall optimization approach is to update β, then update the ancillary parameter estimates, and continue alternating between the estimating equations throughout the iterative optimization.

2.5 Summary

We illustrated three derivations of estimating equations for likelihood-based models with independent data and then showed the relationship of the GLM estimating equation to the previously illustrated models. We discussed the ability of an analyst to build models that extend the GLM to quasilikelihood models. We then introduced the concept of panel data and showed examples of how likelihood-based models are derived to address the correlated nature of the data. We also showed that a naive pooled estimator could be estimated along with a modified sandwich estimate of variance in order to adjust the standard errors of the naive point estimates. Finally, we gave a general overview of how estimation proceeds once an estimating equation is specified.

The middle of the sandwich estimate of variance involves the sums of the contributions in each panel. The use of sums over correlated panels results in a variance estimate called the modified sandwich estimate of variance. See Carroll and Kauermann (*to appear*) and Hardin and Hilbe (2001) for lucid discussions of the robust variance estimate.

Our illustration of deriving estimating equations for likelihood-based models included models for independent data and panel data. Pooled models, unconditional fixed-effects models, conditional fixed-effects models, and random-effects models all admit estimating equations through the same construction algorithm shown for the independent data models.

The following chapter presents the details and motivation of generalized estimating equations. The motivation and illustrations extend the results shown in this chapter. Thus, this review serves to provide the basis of the various kinds of GEE models addressed. You should have a thorough understanding of the techniques, derivations, and assumptions that are presented here in order to fully appreciate the extensions covered in the subsequent material.

2.6 Exercises

1. Show that the negative binomial regression model is a member of the exponential family and discuss ways to address the ancillary parameter. The negative binomial density is given by

$$f(y; r, p) = \binom{y + r - 1}{r - 1} p^r (1-p)^y$$

   where the density provides the probability of observing y failures before the rth success in a series of Bernoulli trials, each with probability of success equal to p.

2. Derive the FIML estimating equation for a binomial regression model. You should be able to incorporate the repeated (Bernoulli) trial nature of the distribution into the earlier Bernoulli example.

3. Derive the FIML estimating equation for the gamma regression model and identify the canonical link function.

4. Derive the conditional fixed-effects linear regression estimating equation.

5. The FIML Poisson model used a log-link for estimating the parameters. Show that the interpretation of the exponentiated coefficients do not depend on the value of the covariate and use the delta method to derive the variance of the natural (not parameterized or *untransformed*) coefficient.

6. Discuss possible parameterizations for the LIML estimating equation, treating σ^2 as ancillary, of an inverse Gaussian(μ, σ^2) model, where the inverse Gaussian density is given by

$$f(y; \mu; \sigma^2) = \frac{1}{\sqrt{2\pi y^3 \sigma^2}} \exp\left\{ -\frac{(y-\mu)^2}{2(\mu\sigma)^2 y} \right\}$$

7. A Gaussian random-effects linear regression model may be derived such that there is an analytic solution to the integral in the log-likelihood. Show this derivation.

8. Discuss the advantages and disadvantages of the pooled estimator for panel data.

9. Give a detailed argument for how to treat the following (complete) panel of data in a conditional fixed-effects logistic regression:

id	y	x1	x2	x3
4	1	0	1	0
4	1	1	1	0
4	1	1	0	1
4	1	0	0	1

10. A conditional fixed-effects model does not, in general, include a parameter for the constant. Show that the conditional fixed-effects negative binomial model does allow a constant in the model and discuss why this is so.

CHAPTER 3

Generalized Estimating Equations

In the previous chapter we illustrated a number of estimating equations that were all derived from a log-likelihood. We showed that the LIMQL estimating equation for GLMs has its genesis in a log-likelihood based upon the exponential family of distributions. In addition, we noted that the utility of this estimating equation is extended outside of the implied log-likelihood due to the work of Wedderburn (1974). The estimating equation methods are related to the quasilikelihood methods in that there are no parametric assumptions.

The term *generalized estimating equations* indicates that an estimating equation is not the result of a likelihood-based derivation, but that it is obtained by generalizing another estimating equation. The modification we make to obtain a generalized estimating equation (GEE) is an introduction of second order variance components directly into a pooled estimating equation. As we saw in the latter sections of the previous chapter the likelihood-based approach would address these additional variance components parametrically. Here, the approach is *ad hoc*.

3.1 Population-averaged (PA) and subject-specific (SS) models

To highlight two different categories of models, let us consider the generalized linear mixed model as the source of nonindependence. For a given outcome y_{it}, we have a $(p \times 1)$ vector of covariates $\mathbf{x}_{it}$ associated with our parameter vector $\boldsymbol{\beta}$. We also have a $(q \times 1)$ vector of covariates $\mathbf{z}_{it}$ associated with the random effect $\boldsymbol{\nu}_i$. The conditional expectation of the outcome is given by

$$\mu_{it}^{\text{SS}} = \text{E}(y_{it}|\boldsymbol{\nu}_i) \qquad (3.1)$$

The responses for a given panel i are characterized by

$$g(\mu_{it}^{\text{SS}}) = \mathbf{x}_{it}\boldsymbol{\beta}^{\text{SS}} + \mathbf{z}_{it}\boldsymbol{\nu}_i \qquad (3.2)$$
$$\text{V}(y_{it}|\boldsymbol{\nu}_i) = \text{V}(\mu_{it}^{\text{SS}}) \qquad (3.3)$$

where the random effects $\boldsymbol{\nu}_i$ follow some distribution.

We can either focus on the distribution of the random effects as the source of nonindependence, or we can consider the marginal expectation of the outcome (integrated over the distribution)

$$\mu_{it}^{\text{PA}} = \text{E}\big[\text{E}(y_{it}|\boldsymbol{\nu}_i)\big] \qquad (3.4)$$

so that the responses are characterized by

$$g(\mu_{it}^{PA}) = \mathbf{x}_{it}\beta^{PA} \quad (3.5)$$
$$V(y_{it}) = V(\mu_{it}^{PA})a(\phi) \quad (3.6)$$

Thus, the marginal expectation is the average response for observations sharing the same covariates (across all panels).

The SS and PA superscripts are added above to differentiate the two approaches. SS indicates that we are explicitly modeling the source of the heterogeneity and that the coefficients β^{SS} have an interpretation for individuals—the SS means *subject specific*. PA indicates that we are looking at the marginal outcome averaged over the population of individuals and that the coefficients β^{PA} have an interpretation in terms of the response averaged over the population—the PA means *population averaged*. One should also note that the form of the marginal model is a parameterization in terms of the distribution of the panels. As such, variance weighted analyses are limited to include weights that are at the level of the panel (and not at level of the observation).

Likelihood-based models that fit in these categories include random effects probit regression models (subject specific) and beta-binomial marginal regression models (population averaged).

Sribney (1999) devised an illustrative example highlighting the difference between the parameters for SS and PA models. At the heart of the illustration is an emphasis that the population parameters β^{SS} and β^{PA} are different entities. The SS model fully parameterizes the distribution of the population, while the PA model parameterizes only the marginal distribution of the population.

Suppose that we are entertaining a logistic regression model where the outcome of interest Y_{it} represents the case of a child having a respiratory illness (1/0). A single explanatory covariate X_{it} denotes the smoking status of the child's mother. The SS model with a single random component ν_i assumes that

$$\text{logit } P(Y_{it} = 1 | X_{it}, \nu_i) = \beta_0^{SS} + X_{it}\beta_1^{SS} + \nu_i \quad (3.7)$$

such that the subject-specific odds ratio given by

$$\text{OR}^{SS} = \frac{P(Y_{it} = 1 | X_{it} = 1, \nu_i)/P(Y_{it} = 0 | X_{it} = 1, \nu_i)}{P(Y_{it} = 1 | X_{it} = 0, \nu_i)/P(Y_{it} = 0 | X_{it} = 0, \nu_i)} = \exp(\beta^{SS}) \quad (3.8)$$

represents the ratio of the odds of a given child having respiratory illness if the mother smokes compared to the odds of the *same* child having respiratory illness if the mother does not smoke.

The PA model, on the other hand, assumes that

$$\text{logit } P(Y_{it} = 1 | X_{it}) = \beta_0^{PA} + X_{it}\beta_1^{PA} \quad (3.9)$$

such that the population-averaged odds ratio given by

$$\text{OR}^{PA} = \frac{P(Y_{it} = 1 | X_{it} = 1)/P(Y_{it} = 0 | X_{it} = 1)}{P(Y_{it} = 1 | X_{it} = 0)/P(Y_{it} = 0 | X_{it} = 0)} = \exp(\beta^{PA}) \quad (3.10)$$

represents the ratio of the odds of an average child with respiratory illness and a smoking mother to the odds of an average child with respiratory illness and a nonsmoking mother.

The lesson here is that we must think carefully about which parameter we are interested in estimating. If we wish to estimate how cessation of smoking might decrease the chances of our children getting respiratory illness, we want to estimate a subject-specific model. If we wish to compare the respiratory illness for children of smokers to children of nonsmokers, then we want to estimate a population-averaged model.

3.2 The PA-GEE for GLMs

Certainly the most well-known GEE-derived group of models is that collection described in the landmark paper of Liang and Zeger (1986). The authors therein provide the first introduction to generalized estimating equations. They also provide the theoretical justification and asymptotic properties for the resulting estimators. In fact, the majority of researchers who refer to a GEE model are referring to this particular collection of models.

Understanding the PA-GEE is relatively straightforward given our focus on the development of the estimating equation in the preceding chapter. We begin with the LIMQL estimating equation for GLMs

$$\Psi(\beta) = \left[\left\{ \sum_{i=1}^{n} \sum_{t=1}^{n_i} \frac{y_{it} - \mu_{it}}{a(\phi) V(\mu_{it})} \left(\frac{\partial \mu}{\partial \eta} \right)_{it} x_{jit} \right\}_{j=1,\ldots,p} \right]_{p \times 1} = [\mathbf{0}]_{p \times 1} \quad (3.11)$$

and rewrite it in matrix terms of the panels

$$\Psi(\beta) = \left[\left\{ \sum_{i=1}^{n} \mathbf{x}_{ji}^{\mathrm{T}} \mathrm{D} \left(\frac{\partial \mu}{\partial \eta} \right) [\mathrm{V}(\boldsymbol{\mu}_i)]^{-1} \left(\frac{\mathbf{y}_i - \boldsymbol{\mu}_i}{a(\phi)} \right) \right\}_{j=1,\ldots,p} \right]_{p \times 1}$$

$$= [\mathbf{0}]_{p \times 1} \quad (3.12)$$

where D() denotes a diagonal matrix. $\mathrm{V}(\boldsymbol{\mu}_i)$ is clearly a diagonal matrix which can be decomposed into

$$\mathrm{V}(\boldsymbol{\mu}_i) = \left[\mathrm{D}(\mathrm{V}(\mu_{it}))^{1/2} \, \mathrm{I}_{(n_i \times n_i)} \, \mathrm{D}(\mathrm{V}(\mu_{it}))^{1/2} \right]_{n_i \times n_i} \quad (3.13)$$

This presentation makes it clear that the estimating equation is treating each observation within a panel as independent. A (pooled) model associated with this estimating equation is called the *independence model*.

If we focus on the marginal distribution of the outcome, for which the expected value and variance functions are averaged over the panels (they are unchanged from the specification given for the LIMQL estimating equation for GLMs), then the identity matrix in equation 3.13 is clearly the within-panel correlation matrix. The GEE proposed by Liang and Zeger is a modification of the LIMQL estimating equation for GLMs that simply replaces the identity matrix with a more general correlation matrix, since the variance matrix for correlated data does not have a diagonal form.

$$V(\boldsymbol{\mu}_i) = \left[D(V(\mu_{it}))^{1/2} \, \mathbf{R}(\boldsymbol{\alpha})_{(n_i \times n_i)} \, D(V(\mu_{it}))^{1/2} \right]_{n_i \times n_i} \quad (3.14)$$

We write $\mathbf{R}(\boldsymbol{\alpha})$ to emphasize that the correlation matrix is to be estimated through the parameter vector $\boldsymbol{\alpha}$. It turns out that it is relatively easy to describe a large grouping of useful structures on the correlation matrix via the $\boldsymbol{\alpha}$ parameter vector.

The conceptual idea behind these GEEs is simple. However, that is not to say that the proof is simple, or that our appreciation of the results is any way lessened. Indeed, the Liang and Zeger paper is an impressive presentation of sophisticated asymptotic and distributional statistics. Since our focus in this text is on the concepts and application, we forego the advanced mathematics required to prove the properties of the estimators for these models.

3.2.1 Parameterizing the working correlation matrix

We gain efficiency in the estimation of the regression parameters by choosing to formally include a hypothesized structure to the within-panel correlation. There are several ways in which we might hypothesize this structure.

Only one additional scalar parameter need be estimated if we believe that the observations within a panel follow no specific order and that they are equally correlated. Alternatively, we may hypothesize a more complicated structure under the belief that the observations within a panel do follow a specific order. Here, we may require a vector of additional parameters requiring up to $(\max_i n_i) - 1$ parameters, or an entire matrix of parameters requiring up to $\binom{\max_i n_i}{2} - n_i$ additional parameters.

The following subsections present standard approaches to specifying a structure for the estimated within panel correlation. In each subsection we include the results for analyzing a Poisson model of the repeated observations of seizures for a group of epileptics. The observations are part of the Progabide dataset given in section 5.2.3.

We can formally write the estimator for the ancillary association parameters as the estimating equation

$$\Psi(\boldsymbol{\alpha}) = \sum_{i=1}^{n} \left(\frac{\partial \boldsymbol{\xi}_i}{\partial \boldsymbol{\alpha}} \right)^{\mathrm{T}} \mathbf{H}_i^{-1} (\mathbf{W}_i - \boldsymbol{\xi}_i) = [\mathbf{0}]_{q \times 1} \quad (3.15)$$

where

$$\mathbf{W}_i = (r_{i1}r_{i2}, r_{i1}r_{i3}, \ldots, r_{in_i-1}r_{in_i})^{\mathrm{T}}_{q \times 1} \quad (3.16)$$
$$\mathbf{H}_i = \mathrm{D}\left(V(\mathbf{W}_{ij})\right)_{q \times q} \quad (3.17)$$
$$\boldsymbol{\xi}_i = \mathrm{E}(\mathbf{W}_i)_{q \times 1} \quad (3.18)$$

such that r_{ij} is the ijth Pearson residual, $\mathbf{H}_i$ is a diagonal matrix, and $q = \binom{n_i}{2}$. From this estimating equation, it is clear that the parameterization of the correlation matrix enters through equation 3.18. In fitting this estimating equation, we substitute $\widehat{r}_{ij}$ obtained from the current estimate $\widehat{\beta}$, for the Pearson residuals. In the subsections to follow, we include simple formulae for the estimation of the components of $\boldsymbol{\alpha}$. In all cases the formulae may be derived directly from the above estimating equation.

In the following subsections, symmetric (square) matrices print results only for the lower triangle of the matrix for ease of readability.

3.2.1.1 Exchangeable correlation

The simplest form of the correlation matrix is the identity matrix assumed by the independence model, which imposes no additional ancillary parameters. In a simple extension to this structure, we might hypothesize that observations within a panel have some common correlation (one additional ancillary parameter). In this case, $\boldsymbol{\alpha}$ is a scalar and the working correlation matrix has the structure

$$\mathbf{R}(\boldsymbol{\alpha}) = \begin{bmatrix} 1 & \alpha & \alpha & \cdots & \alpha \\ \alpha & 1 & \alpha & \cdots & \alpha \\ \alpha & \alpha & 1 & \cdots & \alpha \\ \vdots & \vdots & \vdots & \ddots & \vdots \\ \alpha & \alpha & \alpha & \cdots & 1 \end{bmatrix} \quad (3.19)$$

which we can write succinctly as

$$R_{uv} = \begin{cases} 1 & \text{if } u = v \\ \alpha & \text{otherwise} \end{cases} \quad (3.20)$$

This hypothesis is valid for datasets in which the repeated measurements have no time dependence and any permutation of the repeated measurements is valid. An example of this type of data is a health study in which the panels represent clinics and the repeated measurements are patients within the clinics.

A GEE with an exchangeable correlation structure uses the estimated Pearson residuals $\widehat{r}_{it} = (y_{it} - \widehat{\mu}_{it})/\sqrt{V(\widehat{\mu}_{it})}$ from the current fit of the model to estimate the common correlation parameter. The estimate of α using these residuals is

$$\widehat{\alpha} = \frac{1}{\widehat{\phi}} \sum_{i=1}^{n} \left\{ \frac{\sum_{u=1}^{n_i} \sum_{v=1}^{n_i} \widehat{r}_{iu}\widehat{r}_{iv} - \sum_{u=1}^{n_i} \widehat{r}_{iu}^2}{n_i(n_i - 1)} \right\} \quad (3.21)$$

This type of correlation goes under several names including *exchangeable correlation, equal correlation, common correlation,* and *compound symmetry.*

Specifying this structure on the working correlation matrix, the result of fitting this model to the seizure data (section 5.2.3) is given by

```
GEE population-averaged model                Number of obs     =      295
Group variable:                        id    Number of groups  =       59
Link:                                 log    Obs per group: min =        5
Family:                           Poisson                   avg =      5.0
Correlation:                 exchangeable                   max =        5
                                              Wald chi2(3)    =     0.92
Scale parameter:                        1     Prob > chi2     =   0.8203

                         (standard errors adjusted for clustering on id)
------------------------------------------------------------------------
             |              Semi-robust
    seizures |     Coef.    Std. Err.      z    P>|z|    [95% Conf. Interval]
-------------+----------------------------------------------------------
        time |   .111836    .1169256    0.96   0.339    -.1173339    .3410059
    progabide|   .0275345   .2236916    0.12   0.902    -.410893     .465962
    timeXprog|  -.1047258   .2152769   -0.49   0.627    -.5266608    .3172092
       _cons |   1.347609   .1587079    8.49   0.000     1.036547    1.658671
    lnPeriod |  (offset)
------------------------------------------------------------------------
```

The estimated correlation matrix is

```
        c1      c2      c3      c4      c5
r1  1.0000
r2  0.7767  1.0000
r3  0.7767  0.7767  1.0000
r4  0.7767  0.7767  0.7767  1.0000
r5  0.7767  0.7767  0.7767  0.7767  1.0000
```

Note that all off-diagonal correlations are identical, which is characteristic of the exchangeable correlation structure.

This type of correlation can come about when we believe that the repeated measures are not in any particular order. For this particular dataset, we would have to believe that there was no time dependence of the observations despite the fact that the observations are, in fact, collected over time. One could argue in this case that the analysis would benefit from hypothesizing a time series model of the correlation.

In this subsection we illustrate the estimation techniques for fitting the PA-GEE model. To clearly demonstrate the algorithm, we use the following data so that the calculations can be understood and verified by the reader.

id	t	y	x
1	1	4	0
1	2	5	1
1	3	6	0
1	4	7	1
2	1	5	0
2	2	6	1
2	3	7	0
2	4	8	1

THE PA-GEE FOR GLMS

Our goal is to fit an exchangeable correlation linear regression model

$$y_{it} = \beta_0 + x_{it}\beta_1 \qquad (3.22)$$

where the panel level variance is given by

$$\mathbf{V}(\boldsymbol{\mu}_i) \begin{pmatrix} 1 & \rho & \rho & \rho \\ \rho & 1 & \rho & \rho \\ \rho & \rho & 1 & \rho \\ \rho & \rho & \rho & 1 \end{pmatrix} \mathbf{V}(\boldsymbol{\mu}_i) \qquad (3.23)$$

Since we are fitting a Gaussian model,

$$\mathbf{V}(\boldsymbol{\mu}_i) = I_{(4\times 4)} \qquad (3.24)$$

The variance function in terms of the mean $\boldsymbol{\mu}$ is 1.0 and the scale parameter ϕ must still be estimated.

The starting value for (β_0, β_1) must be specified. From a linear regression (or an independent correlation, linear PA-GEE), we obtain $(\beta_0, \beta_1) = (5.5, 1)$. Now, we must estimate the dispersion parameter and the common correlation parameter. We obtain the fitted values of the model xb $= 5.5 + $ x and the residuals res $=$ y $-$ xb

id	t	y	x	xb	res
1	1	4	0	5.5	-1.5
1	2	5	1	6.5	-1.5
1	3	6	0	5.5	0.5
1	4	7	1	6.5	0.5
2	1	5	0	5.5	-0.5
2	2	6	1	6.5	-0.5
2	3	7	0	5.5	1.5
2	4	8	1	6.5	1.5

An estimate of the dispersion parameter is calculated using equation 3.48 (we could also use equation 3.47)

$$\widehat{\phi} = \frac{1}{8}\sum_{i=1}^{2}\sum_{t=1}^{4}\text{res}_{it}^2 \qquad (3.25)$$

$$= \frac{1}{8}(2.25 + 2.25 + 0.25 + 0.25 + 0.25 + 0.25 + 2.25 + 2.25) \qquad (3.26)$$

$$= \frac{1}{8}(10) = 1.25 \qquad (3.27)$$

The (common) exchangeable correlation coefficient is estimated by using an equivalent formula for equation 3.21

$$\widehat{\rho} = \widehat{\phi}^{-1}\frac{1}{12}\sum_{i=1}^{2}\sum_{t=1}^{4}\sum_{t'>t}\text{res}_{it}\text{res}_{it'} \qquad (3.28)$$

$$= (.8)\frac{1}{12}\Big\{[-1.5(-1.5+.5+.5)-1.5(0.5+0.5)+0.5(0.5)]+$$

$$[-0.5(-0.5+1.5+1.5)-0.5(1.5+1.5)+1.5(1.5)]\Big\} \qquad (3.29)$$

$$= (0.8)\frac{1}{12}([-.5]+[-.5]) \qquad (3.30)$$

$$= (0.8)\frac{-1}{12} = -\frac{1}{15} = -.06667 \qquad (3.31)$$

The output for fitting this model is displayed as

```
GEE population-averaged model          Number of obs       =        8
Group variable:                id      Number of groups    =        2
Link:                    identity      Obs per group: min  =        4
Family:                  Gaussian                     avg  =      4.0
Correlation:         exchangeable                     max  =        4
                                       Wald chi2(1)        =     1.50
Scale parameter:             1.25      Prob > chi2         =   0.2207
------------------------------------------------------------------------
       y |      Coef.   Std. Err.       z    P>|z|    [95% Conf. Interval]
---------+--------------------------------------------------------------
       x |          1   .8164966     1.22    0.221    -.6003039   2.600304
   _cons |        5.5   .5400617    10.18    0.000     4.441498   6.558502
------------------------------------------------------------------------
```

with estimated correlation matrix

```
         c1         c2         c3         c4
r1   1.0000
r2  -0.0667    1.0000
r3  -0.0667   -0.0667    1.0000
r4  -0.0667   -0.0667   -0.0667    1.0000
```

The output from the software matches the manual calculations that are illustrated. In a later section we discuss the implications of the two referenced equations for estimating the dispersion parameter. Running this example in software that uses equation 3.47 results in $\widehat{\phi} = 1.6667$ and $\rho = -.06$ respectively. The changes to the manual calculations that match such software results are given by

$$\widehat{\phi} = \frac{1}{8-2}(10) = \frac{1}{6}(10) = 1.6667 \qquad (3.32)$$

$$\widehat{\rho} = (.6)\frac{1}{12-2}(-1) = (.6)\frac{-1}{10} = -.0600 \qquad (3.33)$$

THE PA-GEE FOR GLMS   63

For the specific case of Gaussian variance with an identity link, the exchangeable correlation PA-GEE estimates the same model as random effects linear regression. This equivalence for the PA model to the SS random effects linear regression holds since the model specifies the identity link. It turns out that there are a number of equivalent estimators for this model.

The model may be estimated by fitting a FIML random-effects regression, the PA-GEE exchangeable linear regression, or a generalized least squares (GLS) model. Interpretation of coefficients is identical for the three approaches listed, but numeric differences can arise for three different reasons. The first source of numeric differences is in the choice of estimator for the dispersion parameter in the PA-GEE model. A second source of numeric differences is in whether the dataset is balanced ($t_i = T$ for all $i = 1, \ldots, n$). The third source of numeric differences is whether the dataset is large enough to admit reliable estimates in the FIML model.

Many software packages will allow specification of all of these models. While they all estimate the same underlying population parameters, numeric differences will be noted due to differences in the estimation of ancillary parameters as well as the sensitivity of FIML optimization routines in the specific software package.

Using the balanced linear regression data given in section 5.2.5, we first fit an exchangeable PA-GEE model where the dispersion parameter is estimated using equation 3.48.

```
GEE population-averaged model            Number of obs      =        80
Group variable:                   id     Number of groups   =        10
Link:                       identity     Obs per group: min =         8
Family:                     Gaussian                    avg =       8.0
Correlation:            exchangeable                    max =         8
                                         Wald chi2(2)      =     53.74
Scale parameter:            1.029535     Prob > chi2       =    0.0000

------------------------------------------------------------------------
        y |    Coef.   Std. Err.      z    P>|z|    [95% Conf. Interval]
----------+-------------------------------------------------------------
       x1 | 1.182527   .2342497    5.05    0.000    .7234056    1.641648
       x2 | 1.152869   .2685991    4.29    0.000    .6264242    1.679313
    _cons |  .877244   .2477123    3.54    0.000    .3917368    1.362751
----------+-------------------------------------------------------------
      rho |    .0639
------------------------------------------------------------------------
```

Next, we fit an exchangeable PA-GEE model where the dispersion parameter is estimated using equation 3.47.

```
GEE population-averaged model            Number of obs      =        80
Group variable:                   id     Number of groups   =        10
Link:                       identity     Obs per group: min =         8
Family:                     Gaussian                    avg =       8.0
Correlation:            exchangeable                    max =         8
                                         Wald chi2(2)      =     51.63
Scale parameter:            1.069642     Prob > chi2       =    0.0000
```

```
------------------------------------------------------------------------
         y  |      Coef.   Std. Err.       z    P>|z|   [95% Conf. Interval]
-------------+----------------------------------------------------------
        x1  |   1.18206    .2389458     4.95    0.000    .7137348   1.650385
        x2  |   1.152993   .2739036     4.21    0.000    .6161515   1.689834
      _cons |   .8773187   .2522629     3.48    0.001    .3828926   1.371745
-------------+----------------------------------------------------------
       rho  |    .0622
------------------------------------------------------------------------
```

Fitting a FIML Gaussian random-effects linear regression model results in:

```
Random-effects ML regression              Number of obs      =        80
Group variable (i) : id                   Number of groups   =        10

Random effects u_i ~ Gaussian             Obs per group: min =         8
                                                         avg =       8.0
                                                         max =         8

                                          LR chi2(2)         =     40.09
Log likelihood  = -114.21672              Prob > chi2        =    0.0000

------------------------------------------------------------------------
         y  |      Coef.   Std. Err.       z    P>|z|   [95% Conf. Interval]
-------------+----------------------------------------------------------
        x1  |   1.182527   .2352395     5.03    0.000    .7214657   1.643587
        x2  |   1.152869   .2686589     4.29    0.000    .6263071    1.67943
      _cons |   .877244    .2477359     3.54    0.000    .3916905   1.362797
-------------+----------------------------------------------------------
    /sigma_u|   .2564913   .1678537     1.53    0.126   -.0724959   .5854785
    /sigma_e|   .9817064   .083012     11.83    0.000    .819006    1.144407
-------------+----------------------------------------------------------
       rho  |   .0639005   .0813078                      .0025834   .4015966
------------------------------------------------------------------------
Likelihood ratio test of sigma_u=0: chibar2(01)=    0.92 Prob>=chibar2 = 0.169
```

These results match the results of the first exchangeable PA-GEE model that we fit to this dataset. The results using these two particular estimation approaches do not, in general, match when the data are comprised of unbalanced panels. There can also be small numeric differences for balanced panels when the two estimation approaches use different tolerance criteria for declaring convergence.

The following is the output display of fitting a random effects linear regression model via generalized least squares:

```
Random-effects GLS regression             Number of obs      =        80
Group variable (i) : id                   Number of groups   =        10

R-sq:  within  = 0.4325                   Obs per group: min =         8
       between = 0.0760                                  avg =       8.0
       overall = 0.3871                                  max =         8

Random effects u_i ~ Gaussian             Wald chi2(2)       =     52.88
corr(u_i, X)   = 0 (assumed)              Prob > chi2        =    0.0000
```

```
        y |      Coef.   Std. Err.       z    P>|z|     [95% Conf. Interval]
----------+------------------------------------------------------------------
       x1 |   1.190988    .2367737     5.03   0.000     .7269203    1.655056
       x2 |   1.150578    .2729453     4.22   0.000     .6156148    1.685541
    _cons |   .8759236    .2592818     3.38   0.001     .3677406    1.384107
----------+------------------------------------------------------------------
  sigma_u |  .33430544
  sigma_e |  .99576555
      rho |  .10129539   (fraction of variance due to u_i)
-----------------------------------------------------------------------------
```

Aside from the comparisons that can be made for the estimation methods involved in fitting this same model in various ways, we also point out that the FIML model provides point estimates and standard errors for the variance components since it is estimating all parameters simultaneously. The other methods treat the random effects variance parameters as ancillary.

Hardin and Hilbe (2001) include a sample analysis of insurance claims data.* Observations are collected on the payout y for car insurance claims given the car group (car1, car2, car3) and vehicle age group (value1, value2, value3).

Additional covariates were created for the interaction of the car and vehicle age group indicators. The data are collected on panels defined by the policy holder's age group. Since the group is a collection of different policy holders rather than repeated observations on the same individual, it is reasonable to assume the exchangeable-correlation structure over a time-related structure. Since the outcomes are positive, the natural modeling choices are gamma and inverse Gaussian. Here, we use the gamma model.

```
GEE population-averaged model                   Number of obs      =      871
Group variable:                          pa     Number of groups   =        8
Link:                            reciprocal     Obs per group: min =       14
Family:                               gamma                    avg =    108.9
Correlation:                   exchangeable                    max =      218
                                                Wald chi2(15)      =   580.49
Scale parameter:                    .0315838    Prob > chi2        =   0.0000

-----------------------------------------------------------------------------
        y |      Coef.   Std. Err.       z    P>|z|     [95% Conf. Interval]
----------+------------------------------------------------------------------
     car1 |   .0040821    .0004547     8.98   0.000     .0031909    .0049733
     car2 |   .0036037    .0004137     8.71   0.000     .0027928    .0044146
     car3 |   .0032887    .0005279     6.23   0.000     .0022539    .0043234
   value1 |  -.0017425    .0002973    -5.86   0.000    -.0023253   -.0011598
   value2 |  -.0013117    .0003021    -4.34   0.000    -.0019038   -.0007196
   value3 |   .0001354    .0003812     0.36   0.723    -.0006117    .0008824
  car1val1|  -.0028558    .0004559    -6.26   0.000    -.0037493   -.0019623
  car1val2|  -.0026487    .0004585    -5.78   0.000    -.0035473   -.0017501
  car1val3|  -.0025891    .0005509    -4.70   0.000    -.0036688   -.0015094
  car2val1|  -.0020648     .000407    -5.07   0.000    -.0028625   -.0012672
  car2val2|  -.0022636    .0004134    -5.47   0.000    -.0030739   -.0014533
```

* net from http://www.stata.com/users/jhardin for users of Stata. Note that for this example we divided the frequency weight by 10: replace number=int(number/10).

```
    car2val3 |  -.0025697     .0004942    -5.20    0.000    -.0035383    -.0016011
    car3val1 |  -.0024019     .0005261    -4.57    0.000    -.0034331    -.0013707
    car3val2 |  -.0025727     .0005327    -4.83    0.000    -.0036168    -.0015286
    car3val3 |  -.0028441     .0006057    -4.70    0.000    -.0040312    -.0016571
       _cons |    .004301     .0003235    13.30    0.000      .003667      .004935
```

The results are interpreted as the rate at which each Pound Sterling is paid for an average claim over an unspecified unit of time. The fitted coefficient on car1, for example, is the change in rate at which a claim is paid for a random car from group 1 versus a random car from some other group. Since this is a marginal model, we cannot discuss the effect of an individual observation.

3.2.1.2 Autoregressive correlation

It may be more reasonable to assume a time dependence for the association if the repeated observations within the panels have a natural order. For example, in a health study we might have panels representing patients with repeated measurements taken over time.

The correlation structure is assumed to be $\text{corr}(y_{it}, y_{it'}) = \alpha^{|t-t'|}$. For normally distributed y_{it}, this is analogous to a continuous time autoregressive (AR) process.

In this case, $\boldsymbol{\alpha}$ is a vector (it was scalar in the preceding subsection) and we estimate the correlations using the Pearson residuals $\widehat{r}_{it} = (y_{it} - \widehat{\mu}_{it})/\sqrt{V(\widehat{\mu}_{it})}$ from the current fit of the model.

$$\widehat{\boldsymbol{\alpha}} = \frac{1}{\widehat{\phi}} \left[\sum_{i=1}^{n} \left(\frac{\sum_{t=1}^{n_i-0} \widehat{r}_{i,t}\widehat{r}_{i,t+0}}{n_i}, \ldots, \frac{\sum_{t=1}^{n_i-k} \widehat{r}_{it}\widehat{r}_{i,t+k}}{n_i}, \right) \right] \quad (3.34)$$

The correlation matrix is then built from the autoregressive structure implied by the AR correlations. An autoregressive process of order k has nonzero autocorrelations for many more than k lags; the matrix is constant along all major diagonals.

Specifying this correlation structure, the fit of the model for the seizure data (section 5.2.3) is given by

```
GEE population-averaged model                    Number of obs       =        295
Group and time vars:              id t           Number of groups    =         59
Link:                               log          Obs per group: min  =          5
Family:                         Poisson                         avg  =        5.0
Correlation:                      AR(2)                         max  =          5
                                                 Wald chi2(3)        =       1.76
Scale parameter:                      1          Prob > chi2         =     0.6243
```

```
                                  (standard errors adjusted for clustering on id)
---------------------------------------------------------------------------------
             |               Semi-robust
    seizures |     Coef.     Std. Err.       z     P>|z|     [95% Conf. Interval]
-------------+-------------------------------------------------------------------
        time |   .1364146    .1039591      1.31    0.189    -.0673415    .3401707
    progabide|   .0105488    .2187692      0.05    0.962    -.4182309    .4393285
   timeXprog |  -.1080133    .2354003     -0.46    0.646    -.5693895    .3533629
       _cons |   1.326599    .1615541      8.21    0.000     1.009958    1.643239
    lnPeriod |   (offset)
---------------------------------------------------------------------------------
```

The estimated correlation matrix is

```
         c1       c2       c3       c4       c5
r1   1.0000
r2   0.8101   1.0000
r3   0.7445   0.8101   1.0000
r4   0.6563   0.7445   0.8101   1.0000
r5   0.5863   0.6563   0.7445   0.8101   1.0000
```

The Progabide data are a collection of repeated measures over time. Given this, it is a natural assumption that the dependence of the observations might be related to a time series type dependence. The GEE autoregressive correlation structure provides such a model. The difficulty in choosing this type of correlation lies in determining the correct order of the autoregressive process. The QIC information criterion (see chapter 4) is useful in helping the analyst to choose between competing hypothesized correlation models.

Lesaffre and Spiessens (2001) investigate the stability of the quadrature-approximated Gaussian random effects logistic regression, a topic we discuss in section 2.3.2.3. The data include a patient identifier number idnr; the treatment group of the patient trt; the time of the measurement time; and an indicator outcome variable y.* Here, we fit a marginal probit PA-GEE model subject to a hypothesized autoregressive correlation structure of order one. The data are unbalanced and five of the patients are excluded from the analysis since they have only one observation.

```
GEE population-averaged model                Number of obs      =       1651
Group and time vars:           idnr visit    Number of groups   =        245
Link:                              probit    Obs per group: min =          2
Family:                          binomial                   avg =        6.7
Correlation:                        AR(1)                   max =          7
                                             Wald chi2(4)       =      79.13
Scale parameter:                        1    Prob > chi2        =     0.0000
```

---

* http://www.blackwellpublishers.co.uk/rss/Readmefiles/lesaffre.htm is a link to the data in the article.

```
------------------------------------------------------------------------
         y |      Coef.   Std. Err.       z    P>|z|    [95% Conf. Interval]
-----------+------------------------------------------------------------
       trt |   .040119    .1500653      0.27   0.789   -.2540036    .3342416
      time |  -.060425    .0309173     -1.95   0.051   -.1210218    .0001719
   timeXtrt|  -.0163527   .024753      -0.66   0.509   -.0648676    .0321622
     visit |  -.0861492   .0564823     -1.53   0.127   -.1968524    .024554
     _cons |  -.3053814   .1409726     -2.17   0.030   -.5816827   -.0290802
-----------+------------------------------------------------------------
    alpha1 |   0.7089
------------------------------------------------------------------------
```

3.2.1.3 Stationary correlation

As an alternative to the time series autocorrelation hypothesis, we may instead hypothesize that correlations exist for some small number of time units. In this hypothesis, we specify a maximum time difference for which observations might be correlated such that the correlation matrix is banded. In this case, α is a vector of the correlations for up to user-specified k lags.

Using the Pearson residuals $\widehat{r}_{it} = (y_{it} - \widehat{\mu}_{it})/\sqrt{V(\widehat{\mu}_{it})}$ from the current fit of the model, we can estimate the vector of correlation parameters α in the same manner as for the autoregressive correlation.

$$\widehat{\alpha} = \frac{1}{\phi} \left[\sum_{i=1}^{n} \left(\frac{\sum_{t=1}^{n_i-0} \widehat{r}_{i,t}\widehat{r}_{i,t+0}}{n_i}, \ldots, \frac{\sum_{t=1}^{n_i-k} \widehat{r}_{i,t}\widehat{r}_{i,t+k}}{n_i}, \right) \right] \quad (3.35)$$

The hypothesized correlation matrix is banded with 1s down the diagonal, $\widehat{\alpha}_1$ down the first band, $\widehat{\alpha}_2$ down the second band, and so forth.

The correlation matrix may be succinctly described as

$$R_{uv} = \begin{cases} \alpha_{|u-v|} & \text{if } |u-v| \leq k \\ 0 & \text{otherwise} \end{cases} \quad (3.36)$$

Specifying this structure for the working correlation matrix results in the following model for the seizure data (section 5.2.3).

```
GEE population-averaged model              Number of obs       =      295
Group and time vars:              id t     Number of groups    =       59
Link:                               log    Obs per group: min  =        5
Family:                         Poisson                   avg  =      5.0
Correlation:             stationary(2)                    max  =        5
                                           Wald chi2(3)        =     0.43
Scale parameter:                    1      Prob > chi2         =   0.9333

                  (standard errors adjusted for clustering on id)
------------------------------------------------------------------------
           |             Semi-robust
  seizures |      Coef.   Std. Err.       z    P>|z|    [95% Conf. Interval]
-----------+------------------------------------------------------------
      time |   .0866246   .1739279     0.50    0.618   -.2542677    .4275169
  progabide|   .0275345   .2236916     0.12    0.902   -.410893     .465962
  timeXprog|  -.1486518   .2506858    -0.59    0.553   -.639987     .3426833
     _cons |  1.347609    .1587079     8.49    0.000    1.036547    1.658671
  lnPeriod |   (offset)
------------------------------------------------------------------------
```

and the estimated correlation matrix is given by

```
        c1      c2      c3      c4      c5
r1  1.0000
r2  0.8152  1.0000
r3  0.7494  0.8152  1.0000
r4  0.0000  0.7494  0.8152  1.0000
r5  0.0000  0.0000  0.7494  0.8152  1.0000
```

Note the partial similarity of this correlation matrix to the one displayed using the autoregressive correlation structure. In particular, note the bands of 0.0 at the lower left, and hence upper right, extremes of the matrix.

The stationary model differs from the autoregressive model in that the correlations are not assumed to be nonzero after the specified order.

Hardin and Hilbe (2001) provide an analysis of length-of-stay data.* The data include the length of hospital stay in days, los; whether the patient is Caucasian, white; urgent admission indicator, type2; emergency admission indicator, type3; and an indicator of whether the patient died, died. Hospital admissions that are neither urgent nor emergencies are deemed elective. The goal is to model the length of stay on the covariates taking into account the correlation of patients with the same insurance provider, provider. The data are from 54 different providers and the panels are unbalanced; the smallest provider has a single observation and the largest has 92.

Here we wish to use the geometric family to describe the model. This family is the discrete correlate to the negative exponential distribution and may be specified as the negative binomial variance function where the ancillary parameter is fixed at 1.0. While the canonical link for the negative binomial variance is the negative binomial link function, most researchers use the log link to facilitate comparisons to the Poisson model. In fact, most applications of the negative binomial models are a means to address overdispersion in the Poisson model. As a side note, we should mention that a negative binomial model employing the canonical link has the unfortunate property of having the ancillary parameter embedded in both the link and variance functions. Estimation, and hence tractibility, is more troublesome than using the natural log link, $\ln(\mu)$.

Here, we fit the log link geometric model subject to a stationary correlation structure with lag 2. Our use of this correlation structure is illustrative for this section. In reality, we would prefer the exchangeable correlation or unstructured correlation given that the panels are made up of different patients. Given our desire to fit the stationary correlation structure with up to 2 lags, the estimation first drops those panels with fewer than 3 observations. The resulting analysis is on 49 providers rather than the 54 represented in our data.

---

* net from http://www.stata.com/users/jhardin for users of Stata.

```
GEE population-averaged model                    Number of obs      =       1487
Group and time vars:           provider t        Number of groups   =         49
Link:                                log         Obs per group: min =          3
Family:             negative binomial(k=1)                      avg =       30.3
Correlation:                 stationary(2)                      max =         92
                                                 Wald chi2(4)      =      59.29
Scale parameter:                         1       Prob > chi2       =     0.0000
------------------------------------------------------------------------------
         los |       IRR   Std. Err.      z    P>|z|     [95% Conf. Interval]
-------------+----------------------------------------------------------------
       white |   .883152    .086767    -1.26   0.206     .728463    1.070689
       type2 |  1.271154    .0933929    3.27   0.001    1.100677    1.468036
       type3 |  2.017643    .2301166    6.15   0.000    1.613477    2.523049
        died |  .7917637    .0457454   -4.04   0.000     .7069945    .8866968
-------------+----------------------------------------------------------------
      alpha1 |    0.0373
      alpha2 |    0.0577
------------------------------------------------------------------------------
```

The output indicates that both urgent and emergency admissions to the hospital significantly increase the length of stay averaged over the providers. We also see that dying results in a shorter stay averaged over providers.

The output is listed in terms of exponentiated coefficients. For Poisson, negative binomial, and geometric models, the exponentiated coefficient is interpreted as the incidence rate ratio. For illustration, note that the inverse link $\exp(\eta)$ is nonlinear and focus on the interpretation of the coefficient β_1 on white.

$$\Delta los_{it} = \exp(\beta_0 + (\text{white}_{it} + 1)\beta_1 + \text{type2}_{it}\beta_2 + \text{type3}_{it}\beta_3)$$
$$- \exp(\beta_0 + \text{white}_{it}\beta_1 + \text{type2}_{it}\beta_2 + \text{type3}_{it}\beta_3) \quad (3.37)$$

Clearly, the effect on the length of stay, by increasing the value of the white covariate, depends on the values of the other covariates. Instead of focusing on the *difference* in the outcome, we can define a different measure based on the incidence rate ratio in the length of stay

$$\text{IRR}_{\text{white}} = \frac{\exp(\beta_0 + (\text{white}_{it} + 1)\beta_1 + \text{type2}_{it}\beta_2 + \text{type3}_{it}\beta_3)}{\exp(\beta_0 + \text{white}_{it}\beta_1 + \text{type2}_{it}\beta_2 + \text{type3}_{it}\beta_3)} \quad (3.38)$$
$$= \exp(\beta_1) \quad (3.39)$$

This well defined ratio has a clear interpretation and does not depend on the values of the covariates. It is a simple transformation of the fitted coefficients. Standard errors for the exponentiated coefficients can be obtained through the delta method; see Feiveson (1999) for a helpful illustration. We use 1 rather than 0 as the null hypothesis in testing our individual coefficients. An incidence rate ratio of 1 indicates no change in the rate while an incidence rate ratio of 2 indicates twice the incidence for an increase of one in the associated coefficient.

3.2.1.4 Nonstationary correlation

In order to formalize a correlation structure where each entry in the working correlation matrix is estimated from the available information, we specify a matrix of parameters $\boldsymbol{\alpha}$.

$$\widehat{\boldsymbol{\alpha}} = \frac{\sum_{i=1}^{n} n_i}{\sum_{i=1}^{n} \sum_{t=1}^{n_i} \widehat{r}_{i,t}^2 / n_i} \mathbf{G} \tag{3.40}$$

$$\mathbf{G} = \begin{pmatrix} g_{1,1}\widehat{r}_{i,1}^2 & g_{1,2}\widehat{r}_{i,1}\widehat{r}_{i,2} & \cdots & g_{1,n_i}\widehat{r}_{i,1}\widehat{r}_{i,n_i} \\ g_{2,1}\widehat{r}_{i,2}\widehat{r}_{i,1} & g_{2,2}\widehat{r}_{i,2}^2 & \cdots & g_{2,n_i}\widehat{r}_{i,2}\widehat{r}_{i,n_i} \\ \vdots & \vdots & \ddots & \vdots \\ g_{n_i,1}\widehat{r}_{i,n_i}\widehat{r}_{i,1} & g_{n_i,2}\widehat{r}_{i,n_i}\widehat{r}_{i,2} & \cdots & g_{n_i,n_i}\widehat{r}_{i,n_i}^2 \end{pmatrix} \tag{3.41}$$

$$g_{u,v} = \left(\sum_{i=1}^{n} I(i,u,v)\right)^{-1} \tag{3.42}$$

$$I(i,u,v) = \begin{cases} 1 & \text{if panel } i \text{ has observations at indexes } u \text{ and } v \\ 0 & \text{otherwise} \end{cases} \tag{3.43}$$

Similar to the stationary structure, the nonstationary correlation matrix uses the estimated correlations for a specified number of bands g for the matrix. The working correlation matrix is specified as

$$R_{uv} = \begin{cases} 1 & \text{if } u = v \\ \alpha_{uv} & \text{if } 0 < |u - v| \leq g \\ 0 & \text{otherwise} \end{cases} \tag{3.44}$$

The nonstationary hypothesis differs from the stationary hypothesis in that the correlation matrix is not assumed to have constant correlations down the diagonals. This estimated correlation matrix is not guaranteed to be invertible and numeric problems may be encountered, especially for unbalanced datasets. This is readily observed by the different amounts of information going into the matrix element estimates.

```
GEE population-averaged model                    Number of obs      =        295
Group and time vars:               id t          Number of groups   =         59
Link:                              log           Obs per group: min =          5
Family:                            Poisson                     avg =        5.0
Correlation:                       nonst                       max =          5
                                                 Wald chi2(3)      =       0.43
Scale parameter:                    1            Prob > chi2       =     0.9333

                             (standard errors adjusted for clustering on id)
-----------------------------------------------------------------------------
             |             Semi-robust
    seizures |      Coef.   Std. Err.      z    P>|z|     [95% Conf. Interval]
-------------+---------------------------------------------------------------
        time |   .0866246   .1739279     0.50   0.618    -.2542677    .4275169
    progabide|   .0275345   .2236916     0.12   0.902    -.410893     .465962
   timeXprog | -.1486518   .2506858    -0.59   0.553    -.639987    .3426833
       _cons |   1.347609   .1587079     8.49   0.000     1.036547    1.658671
    lnPeriod |  (offset)
-----------------------------------------------------------------------------
```

The estimated correlation matrix is

```
        c1      c2      c3      c4      c5
r1  1.0000
r2  0.9892  1.0000
r3  0.7077  0.8394  1.0000
r4  0.0000  0.9865  0.7291  1.0000
r5  0.0000  0.0000  0.5538  0.7031  1.0000
```

Note the relationship of the above correlation matrix to that of the stationary correlation structure. Both have bands of 0.0 at the matrix extremes.

3.2.1.5 Unstructured correlation

The unstructured correlation matrix is the most general of the correlation structures that we discuss. It imposes no structure to the correlation matrix and is equal to the nonstationary matrix for the maximum lag. The working correlation matrix is specified as

$$\mathbf{R} = \boldsymbol{\alpha} \tag{3.45}$$

where $\boldsymbol{\alpha}$ is defined by equation 3.40. Like the nonstationary correlation structure, the estimated correlation matrix is not guaranteed to be invertible and numeric problems may be encountered, especially for unbalanced datasets. Again as for the nonstationary structure, we can see the source of these problems as the different amounts of information going into the calculation of the individual matrix element estimates.

```
GEE population-averaged model              Number of obs     =      295
Group and time vars:             id t      Number of groups  =       59
Link:                             log      Obs per group: min =        5
Family:                       Poisson                     avg =      5.0
Correlation:             unstructured                     max =        5
                                           Wald chi2(3)      =     0.37
Scale parameter:                    1      Prob > chi2       =   0.9464

                                  (standard errors adjusted for clustering on id)
-----------------------------------------------------------------------------
             |              Semi-robust
    seizures |      Coef.   Std. Err.      z    P>|z|     [95% Conf. Interval]
-------------+---------------------------------------------------------------
        time |   .0826525   .1386302     0.60   0.551    -.1890576    .3543626
    progabide |   .0266499    .224251     0.12   0.905    -.4128741    .4661738
   timeXprog | -.1002765   .2137986    -0.47   0.639    -.5193139     .318761
       _cons |   1.335305   .1623308     8.23   0.000     1.017142    1.653467
    lnPeriod |   (offset)
-----------------------------------------------------------------------------
```

The estimated correlation matrix is given by

```
       c1       c2       c3       c4       c5
r1  1.0000
r2  0.9980   1.0000
r3  0.7149   0.8290   1.0000
r4  0.8034   0.9748   0.7230   1.0000
r5  0.6836   0.7987   0.5483   0.6983   1.0000
```

This correlation matrix is unlike those that we have thus far considered. Note the asymmetry of the off-diagonal values.

3.2.1.6 Fixed correlation

A fixed correlation matrix can be imposed if we have knowledge of the structure of the correlation matrix from another source. This approach does not estimate the working correlation at each step, but rather it takes the supplied correlation matrix as given.

Another use for specifying a fixed correlation matrix is to enable estimation of a structure that is not directly supported by an option of a specific software program. This correlation structure is discussed in an example in the following subsection.

3.2.1.7 Free specification

There are many other structures we might hypothesize for our within panel correlation matrix that do not follow the constraints of the discussed forms. For example, we might have a study in which there are patients within doctors (our panels). We might hypothesize that the multiple observations on a given patient are correlated while observations between patients (even within the same doctor) are uncorrelated as a first step in the analysis. None of the previously discussed correlations match such a description.

For a specific example, consider a balanced panel dataset where the panels identify ophthalmologists. Each ophthalmologist reports on a study of treating each eye of four different patients. If the data are ordered such that within the ophthalmologist id we collect data on the left eye and then the right eye of each patient, we can hypothesize a common correlation of data on eyes for individual patients while data across patients are uncorrelated. Such a hypothesized correlation matrix would take the form

$$\mathbf{R} = \begin{bmatrix} 1 & \rho & 0 & 0 & 0 & 0 & 0 & 0 \\ \rho & 1 & 0 & 0 & 0 & 0 & 0 & 0 \\ 0 & 0 & 1 & \rho & 0 & 0 & 0 & 0 \\ 0 & 0 & \rho & 1 & 0 & 0 & 0 & 0 \\ 0 & 0 & 0 & 0 & 1 & \rho & 0 & 0 \\ 0 & 0 & 0 & 0 & \rho & 1 & 0 & 0 \\ 0 & 0 & 0 & 0 & 0 & 0 & 1 & \rho \\ 0 & 0 & 0 & 0 & 0 & 0 & \rho & 1 \end{bmatrix} \quad (3.46)$$

A final analysis might hypothesize that the zeros in the above hypothesized structure are replaced with τ to denote a nonzero, but different, correlation between patients within doctors. If we really believed that within–patient correlation was the only source of correlation in the data, we would simply specify that the patients should be in the i panel data. We may also consider adding a fixed effect for doctors.

We can impose any structure we wish if we have access to software that allows a fixed correlation matrix specification, a specification for limiting the estimation to a single step, and the ability of the user to supply starting values for the regression coefficients. However, to do so requires some programming, or at least repeated calls to the command for fitting the model.

We can proceed by specifying an independent correlation matrix to get starting values for our regression coefficients. We can then obtain the Pearson residuals and estimate a correlation matrix under any constraints we desire. Next, we iterate by supplying our estimated correlation matrix to the software using the fixed correlation matrix specification along with the starting values for the regression coefficients. We limit the estimation to a single step, take the resulting regression coefficients as input to the next step, obtain an updated estimate of the correlation matrix, and iterate.

Our example uses the data listed in section 5.2.5, which follows the above structure. Since the data are constructed, we do not focus on the nature of the data nor on any particular ophthalmological study. The data exist merely to illustrate the techniques for fitting user-specified correlation structures under the current software options.

If we cannot specify a structure, and the options do not allow sufficient control for estimation through a specified correlation matrix, we could specify an unstructured correlation matrix in order to see if a recognizable structure exists. This specification may not lead to convergence. The results below were obtained after specifying a more liberal convergence criterion. The difficulty in fitting this model is not unexpected since we are estimating 3 regression parameters and 56 association parameters from only 80 observations.

```
GEE population-averaged model           Number of obs      =         80
Group and time vars:              id t  Number of groups   =         10
Link:                         identity  Obs per group: min =          8
Family:                       Gaussian                 avg =        8.0
Correlation:              unstructured                 max =          8
                                        Wald chi2(2)       =     212.47
Scale parameter:              1.137088  Prob > chi2        =     0.0000

                   (standard errors adjusted for clustering on id)
------------------------------------------------------------------------
             |              Semi-robust
          y  |     Coef.    Std. Err.      z    P>|z|   [95% Conf. Interval]
-------------+----------------------------------------------------------
         x1  |  1.357834    .1886613    7.20   0.000    .9880643   1.727603
         x2  |  1.857563    .2706295    6.86   0.000    1.327139   2.387987
       _cons |  .2837581    .1964586    1.44   0.149   -.1012936    .6688099
------------------------------------------------------------------------
```

The estimated unstructured correlation matrix is

```
      c1     c2     c3     c4     c5     c6     c7    c8
r1   1.00
r2   1.00   1.00
r3  -0.18  -0.18   1.00
r4  -0.26  -0.23   0.50   1.00
r5  -0.07  -0.18   0.13   0.10   1.00
r6  -0.28  -0.23   0.26   0.24   0.08   1.00
r7   0.38   0.36   0.24  -0.21  -0.25   0.28   1.00
r8  -0.37  -0.35   0.14   0.04   0.03   0.50   0.24  1.00
```

There is no discernible structure to the estimated correlation matrix. However, what we really want is to be able to fit a model using the structure specified by equation 3.46. To do this, we fit an independent model, obtain the residuals, estimate the common correlation of our structure, construct a working correlation matrix, and then specify that constructed matrix and fitted coefficients into another model estimation. In this final command we allow only one iterative step to the regression parameter estimation. This process is repeated until the change in estimated coefficients between successive runs satisfies a predetermined convergence criterion. Using this algorithm, we obtain the following results (for the regression parameters and the common association parameter) for each iteration of this analysis

β_{x1}	β_{x2}	$\beta_{\_cons}$	ρ
1.1584795	1.1589283	0.88134072	0.00000000
1.1387627	1.1181093	0.91961252	0.54273993
1.1387178	1.1180046	0.91970868	0.54414581
1.1387177	1.1180043	0.91970888	0.54414863
1.1387177	1.1180043	0.91970888	0.54414864
1.1387177	1.1180043	0.91970888	0.54414864

The final estimated model results are

```
GEE population-averaged model           Number of obs       =       80
Group and time vars:            id t    Number of groups    =       10
Link:                       identity    Obs per group: min  =        8
Family:                     Gaussian                   avg  =      8.0
Correlation:       fixed (specified)                   max  =        8
                                        Wald chi2(2)        =   112.62
Scale parameter:            1.029858    Prob > chi2         =   0.0000

                    (standard errors adjusted for clustering on id)
------------------------------------------------------------------------
             |              Semi-robust
           y |      Coef.   Std. Err.     z    P>|z|   [95% Conf. Interval]
-------------+----------------------------------------------------------
          x1 |   1.138718    .1862108   6.12   0.000    .7737513   1.503684
          x2 |   1.118004    .3121714   3.58   0.000    .5061597   1.729849
       _cons |   .9197089    .2081197   4.42   0.000    .5118018   1.327616
------------------------------------------------------------------------
```

with estimated correlation matrix

```
     c1    c2    c3    c4    c5    c6    c7    c8
r1  1.00
r2  0.54  1.00
r3  0.00  0.00  1.00
r4  0.00  0.00  0.54  1.00
r5  0.00  0.00  0.00  0.00  1.00
r6  0.00  0.00  0.00  0.00  0.54  1.00
r7  0.00  0.00  0.00  0.00  0.00  0.00  1.00
r8  0.00  0.00  0.00  0.00  0.00  0.00  0.54  1.00
```

These results closely match the theoretical results for the association parameters that we should get given the specifications of the constructed data. Refer to equation 5.4 to compare the estimated correlation parameter with the constructed values.

3.2.2 Estimating the scale variance (dispersion parameter)

The usual estimate of ϕ is given by

$$\widehat{\phi} = \frac{1}{\left(\sum_{i=1}^{n} n_i\right) - p} \sum_{i=1}^{n} \sum_{t=1}^{n_i} \widehat{r}_{it}^2 \qquad (3.47)$$

where $\sum n_i$ is the total number of observations, $\widehat{r}_{it}$ is the itth Pearson residual, and p is the number of covariates in the model. However, Liang and Zeger point out that any consistent estimate of ϕ is admissible. Most software implementations use equation 3.47, but some use

$$\widehat{\phi} = \frac{1}{\sum_{i=1}^{n} n_i} \sum_{i=1}^{n} \sum_{t=1}^{n_i} \widehat{r}_{it}^2 \qquad (3.48)$$

Equation 3.47 has the advantage:

- Model results (independent correlation) *exactly* match GLM results.

Equation 3.48 has the advantage:

- Results are invariant (with any correlation structure) to panel-level replication changes of the dataset.

In other words, if we make an exact copy of our panel dataset (updating the panel identifiers), equation 3.48 results in exactly the same estimates of β (standard errors have a scale change). The use of equation 3.47 fails to produce the same results. The reason is that the numerator (for either estimator) changes by a factor that is related to the number of observations. Only the denominator for equation 3.48 similarly changes.

It is interesting to note that two of the major software packages have each switched the default equation whereby this parameter is estimated. Stata version 5.0 used equation 3.47, but versions 6.0 and higher use equation 3.48. SAS software made the opposite switch. Version 6.12 used equation 3.48, but versions 8.0 and higher use equation 3.47.

Stata users have the option to use both equations. The default calculation for recent versions is equation 3.48; but the PA-GEE modeling command xtgee includes an option nmp for requesting calculation using equation 3.47.

SAS users also have the option to use both equations. The default calculation in recent versions is equation 3.47, but the PA-GEE modeling command PROC GENMOD includes an option V6CORR with the REPEATED statement for requesting calculation by equation 3.48.

3.2.2.1 Independence models

In this subsection we investigate if there is any difference in the two approaches to estimating the dispersion parameter for independence models.

The data to investigate the effects of the competing estimators for the dispersion parameter are only for pedagogical purposes and we make no attempt to identify the nature of the data nor to interpret the results beyond noting the effect on the estimated regression and association parameters. Our first sample dataset, Sample1, is comprised of the following data

id	t	y	x1	x2
1	1	4	0	0
1	2	5	1	0
1	3	6	0	1
1	4	7	1	1
2	1	5	0	0
2	2	6	1	0
2	3	7	0	1
2	4	8	1	1

Sample2 is constructed from Sample1 where we replicate the panels from Sample1. It is a simple replication of the Sample1 panels. This dataset has exactly the same within panel information as in Sample1, but twice the number of panels. Examining the data, you can see that panel 3 (id=3) is the same as panel 1, and panel 4 is the same as panel 2. We have merely added a single copy of each of the original panels in the Sample1 data.

id	t	y	x1	x2
1	1	4	0	0
1	2	5	1	0
1	3	6	0	1
1	4	7	1	1
2	1	5	0	0
2	2	6	1	0
2	3	7	0	1
2	4	8	1	1
3	1	4	0	0
3	2	5	1	0
3	3	6	0	1
3	4	7	1	1
4	1	5	0	0
4	2	6	1	0
4	3	7	0	1
4	4	8	1	1

On the next three pages we present the results for fitting identical independent correlation models to the base (Sample1) and expanded data (Sample2). The aim is to illustrate the effect on the results of the moment estimator for the dispersion parameter.

Before reading those three pages, think about what you expect for the relationship between the fitted coefficients when modeling the two datasets. In addition, consider the relationship of the estimated correlation parameter for the two datasets. The manner in which we estimate the dispersion parameter affects our ability to demonstrate these kinds of relationships, and affects our ability to demonstrate the type of information contained in the data.

The purpose of the following pages is to illustrate the source of differences in output for different software packages. That software will produce different answers for the same analysis is a constant source of confusion usually leading analysts to consider that one of the software packages is producing incorrect results. That is not necessarily the case.

The PA-GEE model is not fully specified and software vendors may choose *any* unbiased estimate for the dispersion parameter. This choice affects comparisons of equivalent analyses across software packages as well as the relationship to other analyses within the same software package. Our detailed examples with the small data provided on this and the preceding page will highlight the effects of the two most common choices for estimating the dispersion parameter. In fact, we know of no software that does not use one of these two estimators.

Fitting a GLM to the datasets yields the following output:

```
Generalized linear models                          No. of obs      =        8
Optimization     : ML: Newton-Raphson              Residual df     =        5
                                                   Scale param     =       .4
Deviance         =             2                   (1/df) Deviance =       .4
Pearson          =             2                   (1/df) Pearson  =       .4

Variance function: V(u) = 1                        [Gaussian]
Link function    : g(u) = u                        [Identity]
Standard errors  : OIM

Log likelihood   = -5.806330821                    AIC             = 2.201583
BIC              = -4.238324625

------------------------------------------------------------------------------
           y |      Coef.   Std. Err.       z    P>|z|     [95% Conf. Interval]
-------------+----------------------------------------------------------------
          x1 |          1   .4472136      2.24   0.025     .1234775   1.876523
          x2 |          2   .4472136      4.47   0.000     1.123477   2.876523
       _cons |        4.5   .3872983     11.62   0.000     3.740909   5.259091
------------------------------------------------------------------------------
```

$\text{Model}^1_{\text{GLM}}$: Linear model for Sample1

```
Generalized linear models                          No. of obs      =       16
Optimization     : ML: Newton-Raphson              Residual df     =       13
                                                   Scale param     =  .3076923
Deviance         =             4                   (1/df) Deviance =  .3076923
Pearson          =             4                   (1/df) Pearson  =  .3076923

Variance function: V(u) = 1                        [Gaussian]
Link function    : g(u) = u                        [Identity]
Standard errors  : OIM

Log likelihood   = -11.61266164                    AIC             = 1.826583
BIC              = -4.317766167

------------------------------------------------------------------------------
           y |      Coef.   Std. Err.       z    P>|z|     [95% Conf. Interval]
-------------+----------------------------------------------------------------
          x1 |          1   .2773501      3.61   0.000     .4564038   1.543596
          x2 |          2   .2773501      7.21   0.000     1.456404   2.543596
       _cons |        4.5   .2401922     18.73   0.000     4.029232   4.970768
------------------------------------------------------------------------------
```

$\text{Model}^2_{\text{GLM}}$: Linear model for Sample2

Note that the coefficient estimates for the two datasets are exactly the same (as we anticipate). The standard errors are scale different since there are twice as many observations in the second dataset. The relationship between the two standard errors is the scale factor

$$\text{SE}_{\text{Model}^2} = \text{SE}_{\text{Model}^1} \sqrt{\frac{n_{\text{Model}^1} - p}{n_{\text{Model}^2} - p}} \qquad (3.49)$$

Fitting an independent PA-GEE model to the two sample datasets using equation 3.48 to estimate ϕ yields the following results:

```
GEE population-averaged model                  Number of obs        =        8
Group variable:                    id          Number of groups     =        2
Link:                        identity          Obs per group: min   =        4
Family:                      Gaussian                         avg   =      4.0
Correlation:              independent                         max   =        4
                                               Wald chi2(2)         =    40.00
Scale parameter:                  .25          Prob > chi2          =   0.0000

Pearson chi2(8):                 2.00          Deviance             =     2.00
Dispersion (Pearson):             .25          Dispersion           =      .25

------------------------------------------------------------------------------
         y |      Coef.   Std. Err.       z    P>|z|     [95% Conf. Interval]
-----------+------------------------------------------------------------------
        x1 |          1   .3535534     2.83    0.005     .3070481    1.692952
        x2 |          2   .3535534     5.66    0.000     1.307048    2.692952
     _cons |        4.5   .3061862    14.70    0.000     3.899886    5.100114
------------------------------------------------------------------------------
```

Model$^1_{\text{GEE}}$: PA-GEE model using equation 3.48 for $\hat{\phi}$ with Sample1

```
GEE population-averaged model                  Number of obs        =       16
Group variable:                    id          Number of groups     =        4
Link:                        identity          Obs per group: min   =        4
Family:                      Gaussian                         avg   =      4.0
Correlation:              independent                         max   =        4
                                               Wald chi2(2)         =    80.00
Scale parameter:                  .25          Prob > chi2          =   0.0000

Pearson chi2(16):                4.00          Deviance             =     4.00
Dispersion (Pearson):             .25          Dispersion           =      .25

------------------------------------------------------------------------------
         y |      Coef.   Std. Err.       z    P>|z|     [95% Conf. Interval]
-----------+------------------------------------------------------------------
        x1 |          1        .25     4.00    0.000     .510009     1.489991
        x2 |          2        .25     8.00    0.000    1.510009     2.489991
     _cons |        4.5   .2165064    20.78    0.000    4.075655     4.924345
------------------------------------------------------------------------------
```

Model$^2_{\text{GEE}}$: PA-GEE model using equation 3.48 for $\hat{\phi}$ with Sample2

The resulting coefficient estimates match the output of the GLM, but the standard errors are different due to the different denominator used by equation 3.48. The relationship of the standard errors for the PA-GEE model to the standard errors of the associated GLM model with the same data is

$$\text{SE}^k_{\text{GEE}} = \text{SE}^k_{\text{GLM}} \sqrt{\frac{n-p}{n}} \qquad (3.50)$$

The relationship between the PA-GEE standard errors of the two datasets is the scale factor

$$\text{SE}_{\text{Model}^2} = \text{SE}_{\text{Model}^1} \sqrt{\frac{n_{\text{Model}^1}}{n_{\text{Model}^2}}} \qquad (3.51)$$

THE PA-GEE FOR GLMS

Fitting an independent PA-GEE model to the two sample datasets using equation 3.47 to estimate ϕ yields the following results:

```
GEE population-averaged model                  Number of obs       =        8
Group variable:                         id     Number of groups    =        2
Link:                             identity     Obs per group: min  =        4
Family:                           Gaussian                    avg  =      4.0
Correlation:                   independent                    max  =        4
                                               Wald chi2(2)        =    25.00
Scale parameter:                        .4     Prob > chi2         =   0.0000

Pearson chi2(5):                      2.00     Deviance            =     2.00
Dispersion (Pearson):                   .4     Dispersion          =       .4

------------------------------------------------------------------------------
         y |      Coef.   Std. Err.      z    P>|z|     [95% Conf. Interval]
-----------+------------------------------------------------------------------
        x1 |          1   .4472136     2.24   0.025     .1234775    1.876523
        x2 |          2   .4472136     4.47   0.000     1.123477    2.876523
     _cons |        4.5   .3872983    11.62   0.000     3.740909    5.259091
------------------------------------------------------------------------------
```

Model$^1_{GEE}$: PA-GEE model using equation 3.47 for $\hat{\phi}$ with Sample1

```
GEE population-averaged model                  Number of obs       =       16
Group variable:                         id     Number of groups    =        4
Link:                             identity     Obs per group: min  =        4
Family:                           Gaussian                    avg  =      4.0
Correlation:                   independent                    max  =        4
                                               Wald chi2(2)        =    65.00
Scale parameter:                   .3076923    Prob > chi2         =   0.0000

Pearson chi2(13):                     4.00     Deviance            =     4.00
Dispersion (Pearson):             .3076923     Dispersion          = .3076923

------------------------------------------------------------------------------
         y |      Coef.   Std. Err.      z    P>|z|     [95% Conf. Interval]
-----------+------------------------------------------------------------------
        x1 |          1   .2773501     3.61   0.000     .4564038    1.543596
        x2 |          2   .2773501     7.21   0.000     1.456404    2.543596
     _cons |        4.5   .2401922    18.73   0.000     4.029232    4.970768
------------------------------------------------------------------------------
```

Model$^2_{GEE}$: PA-GEE model using equation 3.47 for $\hat{\phi}$ with Sample2

The resulting coefficient estimates and standard errors exactly match the results for the GLM fit of each dataset. Equation 3.47 used in the PA-GEE model to estimate ϕ is the same estimator for the dispersion used to estimate the GLM.

3.2.2.2 Exchangeable models

In this subsection, we continue our investigation of the difference in output resulting from our estimation method for the dispersion parameter with an exchangeable logistic model. Our base sample dataset, Sample3, is comprised of the following data

id	t	y	x1	x2
1	1	0	0	0
1	2	1	1	0
1	3	1	0	1
1	4	0	1	1
2	1	0	1	0
2	2	0	1	0
2	3	0	0	1
2	4	1	1	1

Sample4 is constructed from Sample3 in exactly the same manner in which we expanded Sample1 to construct Sample2. We have merely added a single copy of each of the original panels in the Sample3 data.

id	t	y	x1	x2
1	1	0	0	0
1	2	1	1	0
1	3	1	0	1
1	4	0	1	1
2	1	0	1	0
2	2	0	1	0
2	3	0	0	1
2	4	1	1	1
3	1	0	0	0
3	2	1	1	0
3	3	1	0	1
3	4	0	1	1
4	1	0	1	0
4	2	0	1	0
4	3	0	0	1
4	4	1	1	1

On the next two pages, we present the results for fitting similar exchangeable correlation models to the base and expanded data in order to illustrate the effect on the results of the moment estimator for the dispersion parameter. Before looking at the results of the experiment, we should think carefully about our expectations of the results for fitting models to these two datasets. Do you expect the regression parameters to be the same for the two datasets? Do you expect the association parameters to be the same for analyses on the two datasets? What kind of information was added to the construction of the Sample4 data?

Fitting an exchangeable logistic model to the data yields the following results when we use equation 3.48 to estimate ϕ:

```
GEE population-averaged model              Number of obs      =          8
Group variable:                     id     Number of groups   =          2
Link:                            logit     Obs per group: min =          4
Family:                       binomial                    avg =        4.0
Correlation:              exchangeable                    max =          4
                                           Wald chi2(2)       =       0.44
Scale parameter:                    1      Prob > chi2        =     0.8035

------------------------------------------------------------------------------
         y |      Coef.   Std. Err.      z    P>|z|     [95% Conf. Interval]
-----------+------------------------------------------------------------------
        x1 |  -.3681158   1.599157    -0.23   0.818    -3.502406    2.766174
        x2 |   .9705387   1.723205     0.56   0.573    -2.406882    4.347959
     _cons |  -.7897303   1.525776    -0.52   0.605    -3.780196    2.200736
-----------+------------------------------------------------------------------
     alpha |  -0.2338
------------------------------------------------------------------------------
```

Model$^3_{\text{GEE}}$: PA-GEE exchangeable logistic model using equation 3.48 for $\hat{\phi}$ in Sample3

```
GEE population-averaged model              Number of obs      =         16
Group variable:                     id     Number of groups   =          4
Link:                            logit     Obs per group: min =          4
Family:                       binomial                    avg =        4.0
Correlation:              exchangeable                    max =          4
                                           Wald chi2(2)       =       0.87
Scale parameter:                    1      Prob > chi2        =     0.6457

------------------------------------------------------------------------------
         y |      Coef.   Std. Err.      z    P>|z|     [95% Conf. Interval]
-----------+------------------------------------------------------------------
        x1 |  -.3681158   1.130775    -0.33   0.745    -2.584393    1.848162
        x2 |   .9705387    1.21849     0.80   0.426    -1.417658    3.358736
     _cons |  -.7897303   1.078887    -0.73   0.464    -2.904309    1.324849
-----------+------------------------------------------------------------------
     alpha |  -0.2338
------------------------------------------------------------------------------
```

Model$^4_{\text{GEE}}$: PA-GEE exchangeable logistic model using equation 3.48 for $\hat{\phi}$ in Sample4

The coefficient and correlation estimates exactly match for the two datasets using equation 3.48 to estimate ϕ. The relationship between the PA-GEE standard errors of the two datasets is the scale factor

$$SE_{\text{Model4}} = SE_{\text{Model3}} \sqrt{\frac{n_{\text{Model3}}}{n_{\text{Model4}}}} \tag{3.52}$$

This is the same relationship as for the independence PA-GEE model seen in equation 3.51.

Fitting an exchangeable logistic model to the data yields the following results when we use equation 3.47 to estimate ϕ:

```
GEE population-averaged model              Number of obs      =          8
Group variable:                    id      Number of groups   =          2
Link:                           logit      Obs per group: min =          4
Family:                      binomial                    avg  =        4.0
Correlation:             exchangeable                    max  =          4
                                           Wald chi2(2)       =       0.45
Scale parameter:                 1         Prob > chi2        =     0.7985

------------------------------------------------------------------------------
         y |     Coef.   Std. Err.      z    P>|z|    [95% Conf. Interval]
-----------+------------------------------------------------------------------
        x1 |  .1306614   1.675585     0.08   0.938    -3.153424    3.414747
        x2 |  1.141203   1.718915     0.66   0.507    -2.227808    4.510214
     _cons | -1.20693    1.650726    -0.73   0.465    -4.442293    2.028433
-----------+------------------------------------------------------------------
     alpha | -0.1823
------------------------------------------------------------------------------
```

$\text{Model}^3_{\text{GEE}}$: PA-GEE exchangeable logistic model using equation 3.47 for Sample3

```
GEE population-averaged model              Number of obs      =         16
Group variable:                    id      Number of groups   =          4
Link:                           logit      Obs per group: min =          4
Family:                      binomial                    avg  =        4.0
Correlation:             exchangeable                    max  =          4
                                           Wald chi2(2)       =       0.85
Scale parameter:                 1         Prob > chi2        =     0.6526

------------------------------------------------------------------------------
         y |     Coef.   Std. Err.      z    P>|z|    [95% Conf. Interval]
-----------+------------------------------------------------------------------
        x1 | -.0734883   1.163012    -0.06   0.950    -2.352949    2.205973
        x2 |   1.07385   1.217312     0.88   0.378    -1.312038    3.459738
     _cons | -1.037237   1.128697    -0.92   0.358    -3.249443    1.174969
-----------+------------------------------------------------------------------
     alpha | -0.2083
------------------------------------------------------------------------------
```

$\text{Model}^4_{\text{GEE}}$: PA-GEE exchangeable logistic model using equation 3.47 for Sample4

The coefficient and correlation parameter estimates do not match when using the estimator for ϕ given in equation 3.47. The relationship of the standard errors for the two models is complicated by the fact that the estimated common correlation is now different.

3.2.3 Estimating the PA-GEE model

The first software implementation was given by Karim and Zeger (1989) shortly after the appearance of the initial paper describing the PA-GEE collection of models. The authors provided a macro for use with the SAS software system. In addition to this macro, a standalone C-language source code program was developed by Vince Carey estimating these models for balanced panels. Carey later developed code for fitting alternating logistic regression PA-GEE models. He subsequently developed the YAGS software in addition to C++ code classes for programmers.* Once this code was available, support software was developed for use with many other software packages.

Combining the estimating equations for the regression parameters (equation 3.12) and the ancillary parameters (equation 3.15), the complete PA-GEE is given by

$$\Psi(\boldsymbol{\beta}, \boldsymbol{\alpha}) = (\Psi_\beta(\boldsymbol{\beta}, \boldsymbol{\alpha}), \Psi_\alpha(\boldsymbol{\beta}, \boldsymbol{\alpha})) \tag{3.53}$$

$$= \begin{pmatrix} \sum_{i=1}^{n} \mathbf{x}_{ji}^{\mathrm{T}} \mathrm{D} \left(\frac{\partial \boldsymbol{\mu}_i}{\partial \boldsymbol{\eta}} \right) (\mathrm{V}(\boldsymbol{\mu}_i))^{-1} \left(\frac{\mathbf{y}_i - \boldsymbol{\mu}_i}{a(\phi)} \right) \\ \sum_{i=1}^{n} \left(\frac{\partial \boldsymbol{\xi}_i}{\partial \boldsymbol{\alpha}} \right)^{\mathrm{T}} \mathbf{H}_i^{-1} (\mathbf{W}_i - \boldsymbol{\xi}_i) \end{pmatrix} \tag{3.54}$$

$$\mathrm{V}(\boldsymbol{\mu}_i) = \mathrm{D}(\mathrm{V}(\mu_{it}))^{1/2} \mathbf{R}(\boldsymbol{\alpha}) \mathrm{D}(\mathrm{V}(\mu_{it}))^{1/2} \tag{3.55}$$

Estimation assumes that the estimating equation for the correlations is orthogonal to the estimating equation for $\boldsymbol{\beta}$. At each step in the usual GLM algorithm, we first estimate $\mathbf{R}$, and then use it to estimate $\boldsymbol{\beta}$. Convergence is declared when either the change in parameter estimates is less than a set criterion, or the change in the sum of the squared deviances is less than a given criterion. The squared deviance residuals for various distributions from the exponential family are provided in Table 3.1.

While the deviance may be calculated and used as a criterion for declaring convergence in the optimization, it is not usually reported in software. The deviance plays an important part in the inference for GLMs, but does not have the same properties for PA-GEE models, unless the PA-GEE model uses the independent correlation structure. For example, when using a correlation structure other than independence, the deviance could either increase or decrease with the addition of a covariate. Some packages will include GLM type summary statistics including the deviance in the output where these summary statistics are calculated for the independence model. These statistics are useful in calculating other criterion measures as we see in Chapter 4.

Zeger and Liang provide evidence in their early work that even if an incorrect structure is used for the correlation matrix, that only the efficiency of our estimated $\boldsymbol{\beta}$ is affected. This robustness to misspecification of the correlation structure is purchased through the assumption that the estimating equation

---

* http://www.biostat.harvard.edu/~carey currently has links for these packages.

Distribution (variance)	Squared deviance residual $\widehat{d}_i^2$
Gaussian	$(y_i - \widehat{\mu}_i)^2$
Bernoulli	$\begin{cases} -2\ln(1-\widehat{\mu}_i) & \text{if } y_i = 0 \\ -2\ln(\widehat{\mu}_i) & \text{if } y_i = 1 \end{cases}$
Binomial(k)	$\begin{cases} 2k_i \ln\left(\frac{k_i}{k_i - \widehat{\mu}_i}\right) & \text{if } y_i = 0 \\ 2y_i \ln\left(\frac{y_i}{\widehat{\mu}_i}\right) + 2(k_i - y_i)\ln\left(\frac{k_i - y_i}{k_i - \widehat{\mu}_i}\right) & \text{if } 0 < y_i < k_i \\ 2k_i \ln\left(\frac{k_i}{\widehat{\mu}_i}\right) & \text{if } y_i = k_i \end{cases}$
Poisson	$\begin{cases} 2\widehat{\mu}_i & \text{if } y_i = 0 \\ 2\left\{y_i \ln\left(\frac{y_i}{\widehat{\mu}_i}\right) - (y_i - \widehat{\mu}_i)\right\} & \text{otherwise} \end{cases}$
Gamma	$-2\left\{\ln\left(\frac{y_i}{\widehat{\mu}_i}\right) - \frac{y_i - \widehat{\mu}_i}{\widehat{\mu}_i}\right\}$
Inverse Gaussian	$\dfrac{(y_i - \widehat{\mu}_i)^2}{\widehat{\mu}_i^2 y_i}$

Table 3.1 *Squared deviance residuals*

for the regression coefficients is orthogonal to the estimation equation for the correlation coefficients.

We can further protect ourselves from misspecification of the within-panel correlation assumption by employing the modified sandwich estimate of variance for the estimated β. Recall that the modified sandwich estimate of variance is robust to *any* form of within-panel correlation. In this way we gain efficiency in our estimated β if we have the correct form of within-panel correlation, and we are protected from misspecification if we are wrong. Sutradhar and Das (1999) investigate the efficiency of the regression coefficients under misspecification and provide results for some simulation studies.

The modified sandwich estimate of variance for the complete estimating equation is derived from

$$V_{MS} = \mathbf{A}^{-1} \mathbf{B} \mathbf{A}^{-T} \tag{3.56}$$

$$\mathbf{A} = \begin{bmatrix} -\dfrac{\partial \Psi_\beta}{\partial \beta} & -\dfrac{\partial \Psi_\beta}{\partial \alpha} \\ -\dfrac{\partial \Psi_\alpha}{\partial \beta} & -\dfrac{\partial \Psi_\alpha}{\partial \alpha} \end{bmatrix}^{-1} \tag{3.57}$$

$$\mathbf{B} = \sum_{i=1}^{n} \left(\sum_{t=1}^{n_i} \begin{pmatrix} \Psi_{\beta it} \\ \Psi_{\alpha it} \end{pmatrix}\right) \left(\sum_{t=1}^{n_i} \begin{pmatrix} \Psi_{\beta it} \\ \Psi_{\alpha it} \end{pmatrix}\right)^{T} \tag{3.58}$$

However, since we assume that the two estimating equations are orthogonal, we can write

$$A = \begin{bmatrix} -\dfrac{\partial \Psi_\beta}{\partial \beta} & 0 \\ 0 & -\dfrac{\partial \Psi_\alpha}{\partial \alpha} \end{bmatrix}^{-1} = \begin{bmatrix} \left(-\dfrac{\partial \Psi_\beta}{\partial \beta}\right)^{-1} & 0 \\ 0 & \left(-\dfrac{\partial \Psi_\alpha}{\partial \alpha}\right)^{-1} \end{bmatrix} \quad (3.59)$$

This assumption of orthogonality leads to a block diagonal modified sandwich estimate of variance where the upper left entry is given by

$$\left(-\dfrac{\partial \Psi_\beta}{\partial \beta}\right) \left[\sum_{i=1}^{n} \left(\sum_{t=1}^{n_i} \Psi_{\beta it}\right) \left(\sum_{t=1}^{n_i} \Psi_{\beta it}\right)^\mathrm{T}\right] \left(-\dfrac{\partial \Psi_\beta}{\partial \beta}\right)^\mathrm{T} \quad (3.60)$$

and the lower right entry is given by

$$\left(-\dfrac{\partial \Psi_\alpha}{\partial \alpha}\right) \left[\sum_{i=1}^{n} \left(\sum_{t=1}^{n_i} \Psi_{\alpha it}\right) \left(\sum_{t=1}^{n_i} \Psi_{\alpha it}\right)^\mathrm{T}\right] \left(-\dfrac{\partial \Psi_\alpha}{\partial \alpha}\right)^\mathrm{T} \quad (3.61)$$

Because we are interested only in the regression parameters, there is no need to calculate variances for the ancillary parameters since the matrix is block diagonal. The modified sandwich estimate of variance for the regression parameters is the upper $p \times p$ part of V_MS given by equation 3.60 and the modified sandwich estimate of variance for the association parameters is the lower $q \times q$ part of V_MS given in equation 3.61. The variance of the association parameters is not calculated in the approach given in section 3.2, though the formula is valid. The variance of the association parameters is calculated in the ALR approach (section 3.2.5) as well as other GEE models.

We advise that all users specify the modified sandwich estimate of variance with this model. This is called the empirical variance in SAS, the robust variance in S-PLUS, and the semirobust variance in Stata. Stata calls the variance estimate semirobust due to the use of the expected Hessian in the bread (the **A** matrix) of the sandwich variance estimate. The expected Hessian is not robust to the misspecification of the link function. SUDAAN allows user specification of the expected Hessian with option `zeger` or the observed Hessian with option `binder`. The semirobust variance estimate is the same as the robust variance estimate if the canonical link is used for the model, but the output from Stata is still labelled "semirobust." See Hardin and Hilbe (2001) for details and further discussion of robust versus semirobust sandwich estimates of variance.

Recently, Pan (2001b) introduced an alternative estimator for the variance of the outcome. He noted that the usual correction factor for the modified sandwich estimate of variance in PA-GEE models may be written

$$\mathbf{B} = \sum_{i=1}^{n} \mathrm{D}\left(\dfrac{\partial \mu_i}{\partial \eta}\right)^\mathrm{T} \mathrm{V}(\boldsymbol{\mu}_i)^{-\mathrm{T}} \mathrm{Cov}(\mathbf{y}_i) \mathrm{V}(\boldsymbol{\mu}_i)^{-1} \mathrm{D}\left(\dfrac{\partial \mu_i}{\partial \eta}\right) \quad (3.62)$$

to emphasize that the covariance of the outcome is estimated by

$$\text{Cov}(\mathbf{y}_i) = \mathbf{S}_i \mathbf{S}_i^T \tag{3.63}$$
$$\mathbf{S}_i = \mathbf{y}_i - \boldsymbol{\mu}_i \tag{3.64}$$

Pan's alternate formulation changes the covariance of the outcome term to

$$\text{Cov}(\mathbf{y}_i) = \mathbf{A}_i^{1/2} \left(\frac{1}{n} \sum_{i=1}^{n} \mathbf{A}_i^{-1/2} \mathbf{S}_i \mathbf{S}_i^T \mathbf{A}_i^{-1/2} \right) \mathbf{A}_i^{1/2} \tag{3.65}$$

$$\mathbf{A}_i = D\left(\frac{\partial \boldsymbol{\mu}_i}{\partial \eta}\right) \tag{3.66}$$

arguing that the usual estimate is neither consistent nor efficient since it uses data from only one subject.

Early simulation work demonstrated that the modified sandwich estimate of variance resulting from this alternative formulation has a superior performance to the sandwich estimate of variance above in terms of being closer to nominal levels in simulations. When using this new variance estimate, we emphasize that the formulation assumes that the marginal variance of the outcome is modelled correctly, and that there is some common correlation structure for all panels.

Since material on the internet can last far past its useful life, we also discuss another estimation problem even though it no longer exists in commercial software packages. In early software implementations of the PA-GEE model, there was a mistake in the calculation of the association parameters by the Pearson residuals for the exchangeable correlation model in the case where some panels had only a single observation (such panels are called singletons). In the original Liang and Zeger (1986) paper, the scale estimator and exchangeable correlation were correctly specified (using equation 3.47 for the dispersion parameter) as

$$\widehat{\phi}^{-1} = \frac{1}{(\sum_{i=1}^{n} n_i) - p} \sum_{i=1}^{n} \sum_{t=1}^{n_i} \widehat{r}_{it}^2 \tag{3.67}$$

$$\widehat{\alpha} = \widehat{\phi} \sum_{i=1}^{n} \sum_{t=1}^{n_i} \sum_{t'>t} \widehat{r}_{it} \widehat{r}_{it'} \Big/ \left[\sum_{i=1}^{n} .5 n_i (n_i - 1) - p \right] \tag{3.68}$$

However, different formulas were implemented in the first macro program supporting estimation of these models. Subsequent software implementations then copied these (incorrect in the presence of singleton) formulas from the first macro program.

Current software implementations for all of the packages used in this text handle this issue correctly. To verify this, or to test another software implementation, the following data can be modelled as an exchangeable correlation linear regression model.

id	y	x	id	y	x	id	y	x
1	22.5324	0	7	23.2159	0	10	23.5287	273
2	22.1011	0	8	23.4819	0	10	24.5693	416
3	21.6930	0	8	23.1031	242	10	24.0201	616
4	21.3061	0	8	23.6713	382	10	24.6849	806
6	20.2493	0	8	23.2609	551	12	21.1412	0
6	20.3324	230	8	23.7659	718	12	21.8088	225
6	19.6399	406	9	20.4287	0	12	22.8473	400
6	18.6703	593	9	18.9259	234	12	22.1797	595
6	20.9972	770	10	24.1646	0	12	21.7346	771

The correct estimate of the correlation parameter is 0.953, while an incorrect result of 0.748 is reported in flawed implementations. Note that these values use equation 3.47 for estimation of the dispersion parameter (as implied above).

3.2.4 Convergence of the estimation routine

For most data, the estimation routine converges in relatively few iterations. However, there are times when the estimation of the model does not converge. Often this is due to an instability in the estimation of the correlation matrix. A common cause of nonconvergence is that the solution for the correlations iterates between two (or more) possible outcomes.

If we take the Sample1 dataset previously used and try to fit an exchangeable regression model where the dispersion parameter is estimated using equation 3.47, we see that the estimation alternates between the following two estimates of $\Theta = (\beta, \phi)$

$$\Theta_1 = (1, 2, 4.5, .5556) \tag{3.69}$$
$$\Theta_2 = (0, 0, 4.5, 1.000) \tag{3.70}$$

There are two choices that we can take to address this instability in the estimation. One choice is to use the other estimator for the dispersion parameter. The other choice is to specify a different correlation structure. Specifying a different correlation structure explicitly addresses the fact that the data may not support our original specification, while changing the estimator for the dispersion parameter indirectly changes the correlation parameter estimates via the denominator. Either solution fixes the nonconvergence problem seen in this example.

Alternatively, if the model we are analyzing is binomial, we can use the estimation techniques of the following section or we can rely on the one-step estimates—those estimates resulting from a single iteration of the estimation algorithm; see Lipsitz, Fitzmaurice, Orav, and Laird (1994).

3.2.5 ALR: Estimating correlations for binomial models

Carey, Zeger, and Diggle (1993) point out that the Pearson residuals are not a very good choice for the estimation in the special case when we are fitting a binomial model. They offered the alternative approach that is discussed here.

We can write the correlation between a pair of observations in a panel as

$$\text{Corr}(y_{ij}, y_{ik}) = \frac{P(y_{ij} = 1, y_{ik} = 1) - \mu_{ij}\mu_{ik}}{\sqrt{\mu_{ij}(1 - \mu_{ij})\mu_{ik}(1 - \mu_{ik})}} \quad (3.71)$$

and note that the probability that both observations have values of 1 satisfies

$$\max(0, \mu_{ij} + \mu_{ik} - 1) \leq P(y_{ij} = 1, y_{ik} = 1) \leq \min(\mu_{ij}, \mu_{ik}) \quad (3.72)$$

That means that the correlation is constrained to be within some limits that depend on the mean of the data. On the other hand, the *odds ratio* does not have this restriction.

The *odds* is a ratio of the probability of success to the probability of failure. The odds that $y_{ij} = 1$ given that $y_{ik} = 1$ is then

$$\text{Odds}(y_{ij}; y_{ik} = 1) = \frac{P(y_{ij} = 1, y_{ik} = 1)}{P(y_{ij} = 0, y_{ik} = 1)} \quad (3.73)$$

and the odds that $y_{ij} = 1$ given that $y_{ik} = 0$ is

$$\text{Odds}(y_{ij}; y_{ik} = 0) = \frac{P(y_{ij} = 1, y_{ik} = 0)}{P(y_{ij} = 0, y_{ik} = 0)} \quad (3.74)$$

The odds ratio is the ratio of these two odds

$$\text{Odds Ratio}(y_{ij}, y_{ik}) = \psi_{ijk} = \frac{P(y_{ij} = 1, y_{ik} = 1)P(y_{ij} = 0, y_{ik} = 0)}{P(y_{ij} = 0, y_{ik} = 1)P(y_{ij} = 1, y_{ik} = 0)} \quad (3.75)$$

Instead of estimating correlations with Pearson residuals, we can take every pairwise comparison of odds ratios and find the correlation of those measures. In doing so, it is apparent that a method may be derived to obtain the estimated correlation by fitting a logistic regression model to the pairwise odds ratios (at each step of the optimization).

Recall the outline of the PA-GEE estimation from the previous subsection. We are changing the manner in which $\boldsymbol{\alpha}$ is estimated in this approach by specifying an alternate estimating equation for those ancillary parameters. Instead of estimating correlation coefficients from Pearson residuals, we find the odds-ratio estimate for each of the parameters of the specified correlation matrix. In other words, the log odds ratios are used in a logistic regression to estimate the correlation matrix.

The following notation is complicated by the need to address the combinatoric origin of the values that enter the estimating equations.

We let γ_{ijk} denote the log odds ratio between the outcomes y_{ij} and y_{ik} (it is the log of ψ_{ijk} in equation 3.75). We let $\mu_{ij} = P(y_{ij} = 1)$ and $\nu_{ijk} = P(y_{ij} = 1, y_{ik} = 1)$. We then have

$$\text{logit } P(y_{ij} = 1 | y_{ik}) = \gamma_{ijk} y_{ik} + \ln\left(\frac{\mu_{ij} - \nu_{ijk}}{1 - \mu_{ij} - \mu_{ik} + \nu_{ijk}}\right) \quad (3.76)$$

Note that in analyzing the log odds ratio estimates there is an $\binom{n_i}{2}$ vector of values.

Let $\boldsymbol{\zeta}_i$ be the $\binom{n_i}{2}$ vector

$$\zeta_{ijk} = \text{logit}^{-1}\left\{\gamma_{ijk}y_{ik} + \ln\left(\frac{\mu_{ij} - \nu_{ijk}}{1 - \mu_{ij} - \mu_{ik} + \nu_{ijk}}\right)\right\} \quad (3.77)$$

where γ_{ijk} is parameterized as $\mathbf{Z}\boldsymbol{\alpha}$, and $\mathbf{Z}$ is a $(\binom{n_i}{2} \times q)$ known covariate matrix that defines the relationship between pairs of observations in terms of the appropriate elements of $\boldsymbol{\alpha}$.

Overall, this approach involves a second estimating equation such that $\boldsymbol{\Psi}(\Theta) = [\boldsymbol{\Psi}_\beta(\boldsymbol{\beta},\boldsymbol{\alpha}) \quad \boldsymbol{\Psi}_\alpha(\boldsymbol{\beta},\boldsymbol{\alpha})]$ is given by

$$\boldsymbol{\Psi}(\boldsymbol{\beta},\boldsymbol{\alpha})_{((p+q)\times 1)} = (\boldsymbol{\Psi}_\beta(\boldsymbol{\beta},\boldsymbol{\alpha})_{(p\times 1)}, \boldsymbol{\Psi}_\alpha(\boldsymbol{\beta},\boldsymbol{\alpha})_{(q\times 1)}) \quad (3.78)$$

$$= \begin{pmatrix} \sum_{i=1}^{n} \mathbf{x}_{ji}^T D\left(\frac{\partial \boldsymbol{\mu}_i}{\partial \eta}\right)(V(\boldsymbol{\mu}_i))^{-1}\left(\frac{\mathbf{y}_i - \boldsymbol{\mu}_i}{a(\phi)}\right) \\ \sum_{i=1}^{n} \left(\frac{\partial \boldsymbol{\zeta}_i}{\partial \boldsymbol{\alpha}}\right)^T D\left(\zeta_{ijk}(1-\zeta_{ijk})\right)^{-1}\left(\mathbf{y}_i^* - \boldsymbol{\zeta}_i\right) \end{pmatrix} \quad (3.79)$$

$$V(\boldsymbol{\mu}_i) = D(V(\mu_{it}))^{1/2} \mathbf{R}(\boldsymbol{\alpha}) D(V(\mu_{it}))^{1/2} \quad (3.80)$$

where q is the total number of parameters needed for $\boldsymbol{\alpha}$ to represent the desired correlation matrix structure, and $\mathbf{y}_i^*$ is the $(\binom{n_i}{2} \times 1)$ vector constructed from $\mathbf{y}_i$—the construction is such that it matches the indices of $\boldsymbol{\zeta}_i$.

It turns out that using pairwise odds ratios (instead of Pearson residuals) results in two (assumed to be orthogonal) estimating equations which can be efficiently calculated combining a PA-GEE regression-step estimate with a logistic regression. Because of the fact that the algorithm alternates between a PA-GEE step and a logistic regression, it is called *alternating logistic regression* or ALR. SAS users can estimate models with this alternative method for estimating the ancillary parameters via an option on the REPEATED statement in PROC GENMOD. To be clear, the REPEATED statement respectively specifies either CORR= or LOGOR= to specify correlations estimated using Pearson residuals or using log odds ratios.

To summarize, the ALR approach has the following characteristics:
- The approach leads to estimates and standard errors for the regression coefficients and the log odds ratios.
- β and $\boldsymbol{\alpha}$ are assumed to be orthogonal.
- Even though we gain insight into the association parameters, since β and $\boldsymbol{\alpha}$ are assumed to be orthogonal, the estimating equation is still a GEE1.

We first fit an exchangeable correlation PA-GEE model using the data on prenatal care (see section 3.6.3).

```
GEE population-averaged model           Number of obs      =     2449
Group variable:                   mom   Number of groups   =     1558
Link:                           logit   Obs per group: min =        1
Family:                      binomial                  avg =      1.6
Correlation:             exchangeable                  max =        4
                                        Wald chi2(4)       =   237.57
Scale parameter:                    1   Prob > chi2        =   0.0000

                  (standard errors adjusted for clustering on mom)
------------------------------------------------------------------------
            |               Semi-robust
     prenat | Odds Ratio     Std. Err.       z    P>|z|   [95% Conf. Interval]
------------+-----------------------------------------------------------
     indSpa |  .4059843      .053526      -6.84   0.000    .313534    .525695
    husProf |  1.73484       .2837106      3.37   0.001   1.259089   2.390355
     toilet |  3.639483      .5323812      8.83   0.000   2.732288   4.847891
     ssDist |  .9865406      .0024238     -5.52   0.000    .9818014   .9913027
------------------------------------------------------------------------
```

The output in terms of the regression coefficients is

```
                  (standard errors adjusted for clustering on mom)
------------------------------------------------------------------------
            |               Semi-robust
     prenat |      Coef.     Std. Err.       z    P>|z|   [95% Conf. Interval]
------------+-----------------------------------------------------------
     indSpa | -.9014407     .1318425      -6.84   0.000  -1.159847   -.6430341
    husProf |  .5509154     .163537        3.37   0.001    .2303888   .871442
     toilet | 1.291842      .1462793       8.83   0.000   1.005139   1.578544
     ssDist | -.0135508     .0024569      -5.52   0.000   -.0183662  -.0087353
      _cons |  .0862278     .0907927       0.95   0.342   -.0917227   .2641783
------------------------------------------------------------------------
```

This model estimates $\hat{\alpha} = .7812$. Fitting the same exchangeable correlation model using ALR instead of the usual moment estimates from the Pearson residuals results in

```
            Analysis Of GEE Parameter Estimates
              Empirical Standard Error Estimates

                        Standard    95% Confidence
   Parameter  Estimate    Error        Limits            Z      Pr > |Z|

   Intercept    0.0873    0.0908   -0.0906    0.2652    0.96     0.3362
   indspa      -0.9014    0.1318   -1.1597   -0.6430   -6.84    <.0001
   husprof      0.5537    0.1635    0.2332    0.8742    3.39     0.0007
   toilet       1.2934    0.1464    1.0065    1.5803    8.84    <.0001
   ssdist      -0.0135    0.0025   -0.0184   -0.0087   -5.51    <.0001
   Alpha1       4.5251    0.2646    4.0066    5.0437   17.10    <.0001
```

where Alpha1 = 4.5251 is the common (exchangeable) log odds ratio. It is important to emphasize that the two approaches to modeling a common correlation estimate that common correlation on *different* quantities. It may be tempting to convert the results of this approach to the interpretation of the moment estimate of the residuals. We do not do this. If we want to calculate a PA-GEE model based on a correlation structure for the residuals, then we do that directly.

The standard errors in the output are in terms of the sandwich estimate of variance. The estimating equations are solved separately since the two estimating equations are assumed to be orthogonal. In this output, the sandwich estimate of variance has been calculated for *both* of the estimating equations. We do not recommend calculating the naive variance of the estimating equation of the association parameters along with the sandwich estimate of variance for the regression parameters.

Since this model places more emphasis on the estimation of the association parameters, we should consider methods for calculating diagnostics for this part of the estimation problem in addition to our focus on the regression parameters. Methods for calculating leverage and influence are discussed in Chapter 4.

3.2.6 Summary

Liang and Zeger apply the name *population averaged* GEE to emphasize the nature of the generalization of the original estimating equation due to the focus on the marginal distribution.

The PA-GEE for GLMs is best understood by focusing on what we did and what we did not do in deriving the algorithm. First, we did not start with a probability-based model, nor even a likelihood. We used the estimating equation for a pooled model and then generalized the estimation of the panel-level variance through ancillary parameters. There is an implied quasilikelihood from the GEE model which may or may not coincide to a probability-based model.

Second, note that the model was extended assuming a correlation structure that was estimated by combining information across panels. We estimate the ancillary parameters to get a working correlation matrix. Using that matrix (which is applied to each panel), we then estimate the β regression coefficient parameter vector. Thus, we are focusing on the marginal distribution where the panels are summed together after taking into account the correlation. In effect, we are averaging over the panels. The resulting β is called the population-averaged estimator for this reason.

For example, consider a logit model where we model the presence of respiratory illness of a child at each of several doctors visits on whether the mother of a child smokes (also collected at each visit). The estimated parameter for the smoking status of the mother is a measure of the effect of second-hand (maternal) smoke on respiratory illness averaged over all instances of smoking (1/0). This is without regard to a specific mother or child.

We rewrite the PA-GEE more formally as

$$\Psi(\beta) = \sum_{i=1}^{n} \Psi_i \left[\beta, \widehat{\alpha}(\beta, \widehat{\phi}(\beta)), \widehat{\phi}(\beta)\right] \qquad (3.81)$$

in order to emphasize that

- The GEE that we solve uses moment estimates of both ϕ and α.
- The moment estimates for the ancillary parameters depend on β.
- β and α are assumed to be orthogonal.
- There is a SAS macro available that allows one to fit *heteroskedastic* GEEs (in some situations). The extension assumes that $a(\phi) = \phi_i$ such that a different error variance is estimated for each panel. *

We also note that

- The basic structure of the GEE is the independence model—the LIMQL estimating equation for pooled GLMs.
- We do not obtain standard errors for ancillary parameters when using the original models of Liang and Zeger (section 3.2). This is not an issue if the focus of our analysis is only the regression parameters β. ALR or another GEE model is preferred if the focus of our analysis does include the association parameters.
- Using the modified sandwich estimate of variance means that the standard errors for β are robust to misspecification of the assumed correlation structure of α.
- The misspecification of α affects only the efficiency of our regression estimates, but not the consistency.
- The relative gain in using a PA-GEE model over the independence model is relatively small if the panels are small in size. We recommend the independence model when there are less than 30 panels in a dataset.

We alluded to various nuances in the construction of the sandwich estimate of variance. Principally, we noted that the use of the expected Hessian in the construction of the sandwich estimate of variance could be numerically different from the sandwich estimate of variance constructed with the observed Hessian for the cases in which we do not specify the canonical link function. Software implementations may choose either approach. The expected Hessian is typically used (and documented as the method of Fisher scoring) in a straightforward implementation of the IRLS algorithm. The two approaches may also be mixed (as done by default in SAS and optionally in Stata) for the estimation of generalized linear models. SAS users should see the documentation for PROC GENMOD noting the documentation for the EXPECTED and

---

* http://www.statlab.uni-heidelberg.de/statlib/GEE/GEE1/GEE1_202.DOC is the current source of information on the macro at the time of writing this text. The macro was written by Ulrike Grömping.

SCORING options of the MODEL statement. Stata users should consult the documentation for the glm command noting the documentation for the irls and fisher options.

For completeness, we point out that a generalized least squares (GLS) approach can also lead to valid estimates for this class of models. Lipsitz, Laird, and Harrington (1992) present an approach based on generalized least squares. These estimates are asymptotically equivalent to the PA-GEE results and are similar to the one-step PA-GEE approximation—the approximation resulting from only one iteration of the fitting algorithm.

3.3 The SS-GEE for GLMs

The subject-specific versions of GEEs for GLMs also extend the LIMQL estimating equation for pooled GLMs. They have the same origin as the population averaged models. However, we hypothesize that there is some underlying distribution for random effects in the model that serves as the genesis of the within-panel correlation.

As such, there are three items we must address to build models for these GEEs:

1. We must choose a distribution for the random effect.

2. We must derive the expected value which depends on the link function and the distribution of the random effect.

3. We must derive the variance—a function of the usual variance and the random effect.

Recall the forms of the expected value (mean) for the parametric random-effects models from the beginning of our presentation. We must follow this same approach when deriving the expected values for these SS-GEEs.

Whether the calculation of the expected value has an analytic or numeric solution to the integral depends on the choice of the random-effects distribution and the form of the link function (how the expected value is parameterized).

Formally, we have

$$\mu_{it}^{SS} = \mathrm{E}(y) = \mathrm{E}[\mathrm{E}(y_{it}|\nu_i)] = \int f(\nu_i) g^{-1}(\mathbf{x}\boldsymbol{\beta}_i + \nu_i) d\nu_i \qquad (3.82)$$

This formulation assumes a single random effect ν_i. In fact, we may have several random effects such that

$$\mu_{it}^{SS} = \mathrm{E}(y) = \mathrm{E}[\mathrm{E}(y_{it}|\boldsymbol{\nu}_i)] = \int f(\boldsymbol{\nu}_i) g^{-1}(\mathbf{x}\boldsymbol{\beta}_i + \mathbf{z}_{it}\boldsymbol{\nu}_i) d\boldsymbol{\nu}_i \qquad (3.83)$$

where $\mathbf{z}$ is a vector of covariates associated with the random effects, and f is the multivariate density of the random effect vector $\boldsymbol{\nu}_i$.

Focusing on a single random effect ν_i, we emphasize that for the PA-GEEs, we have

$$g(\mu_{it}^{\text{PA}}) = \mathbf{x}_{it}\beta^{\text{PA}} \qquad (3.84)$$
$$\mu_{it}^{\text{PA}} = \text{E}(y_{it}) = g^{-1}(\mathbf{x}_{it}\beta^{\text{PA}}) \qquad (3.85)$$
$$\text{V}^{\text{PA}}(y_{it}) = \text{V}(\mu_{it}^{\text{PA}})a(\phi) \qquad (3.86)$$

and for the SS-GEEs we have

$$g(\mu_{it}^{\text{SS}}) = \mathbf{x}_{it}\beta^{\text{SS}} + \nu_i \qquad (3.87)$$
$$\mu_{it}^{\text{SS}} = \int f(\nu_i) g^{-1}(\mathbf{x}_{it}\beta^{\text{SS}} + \nu_i) d\nu_i \qquad (3.88)$$
$$\text{V}^{\text{SS}}(y_{it}) = \int \left[g^{-1}(\mathbf{x}_{is}\beta^{\text{SS}} + \nu_i) - \mu_{is}^{\text{SS}} \right] \left[g^{-1}(\mathbf{x}_{it}\beta^{\text{SS}} + \nu_i) - \mu_{it}^{\text{SS}} \right] df(\nu_i)$$
$$+ a(\phi) \text{I}(s=t) \int \text{V}(\mu_{it}^{\text{SS}}) df(\nu_i) \qquad (3.89)$$

where f is the distribution of the random effect ν. The variance matrix for the ith subject is defined in terms of the (s,t) entry.

It is important to see that $g^{-1}(\mu_{it}^{\text{PA}})$ is not equal to $g^{-1}(\mu_{it}^{\text{SS}})$ unless $g()$ is the identity function and the expected value of the random effect is zero (as it is for Gaussian random effects).

The SS-GEE for GLMs is estimated using the same LIMQL estimating equation for pooled GLMs but we substitute equation 3.88 for μ_{it} and equation 3.89 for $\text{V}(\mu_{it})$. The difficulty of this depends on the link function and random-effects distribution f. Some, but not all, choices lead to an analytic solution of the integral equations.

3.3.1 Single random-effects

To admit comparison with the population averaged models already covered, we consider a single random effect ν_i following the Gaussian distribution with mean zero and variance σ_ν^2. The expression for the marginal mean is relatively easy to calculate or approximate for the standard link functions. For the identity link,

$$\mu_{it}^{\text{SS}} = \text{E}(y_{it}) = \int g^{-1}(\mathbf{x}_{it}\beta^{\text{SS}} + \nu_i) dF(\nu_i) \qquad (3.90)$$
$$= \int (\mathbf{x}_{it}\beta^{\text{SS}} + \nu_i) \frac{1}{\sqrt{2\pi\sigma_\nu^2}} \exp\left(-\frac{\nu_i^2}{2\sigma_\nu^2}\right) d\nu_i \qquad (3.91)$$
$$= \mathbf{x}_{it}\beta^{\text{SS}} \qquad (3.92)$$

For the log link,

$$\mu_{it}^{SS} = E(y_{it}) = \int g^{-1}(\mathbf{x}_{it}\boldsymbol{\beta}^{SS} + \nu_i)dF(\nu_i) \tag{3.93}$$

$$= \int \exp\left(\mathbf{x}_i\boldsymbol{\beta}^{SS} + \nu_i\right) \frac{1}{\sqrt{2\pi\sigma_\nu^2}} \exp\left(-\frac{\nu_i^2}{2\sigma_\nu^2}\right) d\nu_i \tag{3.94}$$

$$= \exp\left(\mathbf{x}_{it}\boldsymbol{\beta}^{SS} + \frac{\sigma_\nu^2}{2}\right) \tag{3.95}$$

For the probit link,

$$\mu_{it}^{SS} = E(y_{it}) = \int g^{-1}(\mathbf{x}_{it}\boldsymbol{\beta}^{SS} + \nu_i)dF(\nu_i) \tag{3.96}$$

$$= \int \left(\int_{-\infty}^{\mathbf{x}_{it}\boldsymbol{\beta}^{SS}+\nu_i} \frac{1}{\sqrt{2\pi}} e^{-z^2/2} dz\right) \frac{1}{\sqrt{2\pi\sigma_\nu^2}} \exp\left(-\frac{\nu_i^2}{2\sigma_\nu^2}\right) d\nu_i \tag{3.97}$$

$$= \Phi\left(\frac{\mathbf{x}_{it}\boldsymbol{\beta}^{SS}}{\sqrt{1+\sigma_\nu^2}}\right) \tag{3.98}$$

The logit link has no closed form solution, but may be approximated by

$$\mu_{it}^{SS} = E(y_{it}) = \int g^{-1}(\mathbf{x}_{it}\boldsymbol{\beta}^{SS} + \nu_i)dF(\nu_i) \tag{3.99}$$

$$= \int \left(\frac{e^{\mathbf{x}\boldsymbol{\beta}+\nu_i}}{1+e^{\mathbf{x}\boldsymbol{\beta}+\nu_i}}\right) \frac{1}{\sqrt{2\pi\sigma_\nu^2}} \exp\left(-\frac{\nu_i^2}{2\sigma_\nu^2}\right) d\nu_i \tag{3.100}$$

$$\approx \Phi\left(\frac{\mathbf{x}_{it}\boldsymbol{\beta}^{SS}}{\sqrt{1+c^2\sigma_\nu^2}}\right) \tag{3.101}$$

where $c = 16\sqrt{3}/(15\pi)$.

The most important result is that of equation 3.92, showing the equivalence of the identity link parameterization for the subject-specific and population-averaged approaches. This result means that coefficients for models fit with the identity link have both a subject-specific and a population-averaged interpretation. This result is true for any distribution of the random effect for which the expected value is zero.

Ideally, we would be able to derive similar simple expressions for the variance function as well; but this is not true except for the case when the mean is parameterized with the linear link. However, we need only derive an approximation of the variance $\widetilde{V}(\mathbf{y}_i)$, and so consider a Taylor series expansion of equation 3.89 to obtain

$$V(\mathbf{y}_i) \approx V\left[g^{-1}(\mathbf{x}_{it}\boldsymbol{\beta}^{SS}) + \frac{\partial}{\partial \nu_i}g^{-1}(\mathbf{x}_{it}\boldsymbol{\beta}^{SS} + \nu_i)\right]$$
$$+ \phi E\left\{V\left[g^{-1}(\mathbf{x}_{it}\boldsymbol{\beta}^{SS}) + \frac{\partial}{\partial \nu_i}g^{-1}(\mathbf{x}_{it}\boldsymbol{\beta}^{SS})\nu_i\right]\right\} \quad (3.102)$$
$$\approx D\left(\frac{\partial \mu^{SS}}{\partial \eta^{SS}}\right)^2 \sigma_\nu^2 + \phi V(\mu^{SS})JJ^T = \widetilde{V}(\mathbf{y}_i) \quad (3.103)$$

where J is a vector of indicator variables. The first term is a matrix of variance components for the random effect and the second term is a matrix of the dispersion parameter and usual variance for a GLM.

As in the population averaged models, we can use simple moment estimators for the unknown ancillary parameters.

$$\widehat{\sigma}_\nu^2 = \frac{1}{n}\sum_{i=1}^{n}\left[\left(\mathbf{y}_i - \widehat{\boldsymbol{\mu}}_i^{SS}\right)^T\left(\mathbf{y}_i - \widehat{\boldsymbol{\mu}}_i^{SS}\right) - \widehat{\phi}V(\widehat{\boldsymbol{\mu}}_i^{SS})\right] \quad (3.104)$$

$$\widehat{\phi} = \frac{1}{\sum_{i=1}^{n}n_i}\sum_{i=1}^{n}\sum_{t=1}^{n_i}\frac{(y_{it} - \widehat{\mu}_{it}^{SS})^2 - \widehat{\sigma}_\nu^2}{V(\widehat{\mu}_{it}^{SS})} \quad (3.105)$$

3.3.2 Multiple random-effects

Considering single random-effects allows us to more closely compare the resulting models for the SS-GEE and PA-GEE approaches. However, the SS-GEE allows a richer collection of models in that we can consider multiple random effects. For completeness, we vectorize the equations of interest for this set of models. This illustrates the derivation assuming a Gaussian (normal) distribution for the random vector with mean $\mathbf{0}$ and variance matrix $\boldsymbol{\Sigma}_\nu$.

For the identity link and a random vector of length q,

$$\mu_{it}^{SS} = E(y_{it}) = \int g^{-1}(\mathbf{x}_{it}\boldsymbol{\beta}^{SS} + \boldsymbol{\nu}_{it})dF(\boldsymbol{\nu}_{it}) \quad (3.106)$$
$$= \int (\mathbf{x}_{it}\boldsymbol{\beta}^{SS} + \boldsymbol{\nu}_{it})(2\pi)^{-q/2}|\boldsymbol{\Sigma}_\nu|^{-1/2}\exp\left(-\frac{1}{2}\boldsymbol{\nu}_{it}^T\boldsymbol{\Sigma}_\nu^{-1}\boldsymbol{\nu}_{it}\right)d\boldsymbol{\nu}_{it} \quad (3.107)$$
$$= \mathbf{x}_{it}\boldsymbol{\beta}^{SS} \quad (3.108)$$

For the log link,

$$\mu_{it}^{SS} = E(y_{it}) = \int g^{-1}(\mathbf{x}_{it}\boldsymbol{\beta}^{SS} + \boldsymbol{\nu}_{it})dF(\boldsymbol{\nu}_{it}) \quad (3.109)$$
$$= \int e^{(\mathbf{x}_{it}\boldsymbol{\beta}^{SS} + \boldsymbol{\nu}_{it})}(2\pi)^{-q/2}|\boldsymbol{\Sigma}_\nu|^{-1/2}\exp\left(-\frac{1}{2}\boldsymbol{\nu}_{it}^T\boldsymbol{\Sigma}_\nu^{-1}\boldsymbol{\nu}_{it}\right)d\boldsymbol{\nu}_{it} \quad (3.110)$$
$$= \exp\left(\mathbf{x}_{it}\boldsymbol{\beta}^{SS} + \frac{1}{2}\boldsymbol{\nu}_{it}^T\boldsymbol{\Sigma}_\nu\boldsymbol{\nu}_{it}\right) \quad (3.111)$$

For the probit link,

$$\mu_{it}^{SS} = E(y_{it}) = \int g^{-1}(\mathbf{x}_{it}\boldsymbol{\beta}^{SS} + \boldsymbol{\nu}_{it})dF(\boldsymbol{\nu}_{it}) \quad (3.112)$$

$$= \int \left(\int_{-\infty}^{\mathbf{x}_{it}\boldsymbol{\beta}^{SS}+\nu_{it}} (2\pi)^{-q/2}|\mathbf{I}|^{-1/2} \exp\left(-\frac{1}{2}\mathbf{z}_{it}^T\mathbf{I}^{-1}\mathbf{z}_{it}\right) d\mathbf{z}_{it} \right)$$

$$(2\pi)^{-q/2}|\boldsymbol{\Sigma}_\nu|^{-1/2} \exp\left(-\frac{1}{2}\boldsymbol{\nu}_{it}^T\boldsymbol{\Sigma}_\nu^{-1}\boldsymbol{\nu}_{it}\right) d\boldsymbol{\nu}_{it} \quad (3.113)$$

$$= \Phi\left(\mathbf{x}_{it}\boldsymbol{\beta}^{SS}|\boldsymbol{\Sigma}_\nu\boldsymbol{\nu}_{it}\boldsymbol{\nu}_{it}^T + \mathbf{I}|^{-q/2}\right) \quad (3.114)$$

The logit link has no closed form solution, but may be approximated by

$$\mu_{it}^{SS} = E(y_{it}) = \int g^{-1}(\mathbf{x}_{it}\boldsymbol{\beta}^{SS} + \boldsymbol{\nu}_{it})dF(\boldsymbol{\nu}_{it}) \quad (3.115)$$

$$= \int \left(\frac{e^{\mathbf{x}\boldsymbol{\beta}+\nu_{it}}}{1+e^{\mathbf{x}\boldsymbol{\beta}+\nu_{it}}}\right)(2\pi)^{-q/2}|\boldsymbol{\Sigma}_\nu|^{-1/2} \exp\left(-\frac{1}{2}\boldsymbol{\nu}_{it}^T\boldsymbol{\Sigma}_\nu^{-1}\boldsymbol{\nu}_{it}\right)d\boldsymbol{\nu}_{it} \quad (3.116)$$

$$\approx \Phi\left(\mathbf{x}_{it}\boldsymbol{\beta}^{SS}|c^2\boldsymbol{\Sigma}_\nu\boldsymbol{\nu}_{it}\boldsymbol{\nu}_{it}^T + \mathbf{I}|^{-q/2}\right) \quad (3.117)$$

where $c = 16\sqrt{3}/(15\pi)$.

Taking into account the more general matrix notation associated with the random effects, the variance $\tilde{V}(\mathbf{y}_i)$ is derived as

$$V(\mathbf{y}_i) \approx V\left[g^{-1}(\mathbf{x}_{it}\boldsymbol{\beta}^{SS}) + \frac{\partial}{\partial \boldsymbol{\nu}_i}g^{-1}(\mathbf{x}_{it}\boldsymbol{\beta}^{SS} + \boldsymbol{\nu}_i)\right]$$

$$+ \phi E\left\{V\left[g^{-1}(\mathbf{x}_{it}\boldsymbol{\beta}^{SS}) + \frac{\partial}{\partial \boldsymbol{\nu}_i}g^{-1}(\mathbf{x}_{it}\boldsymbol{\beta}^{SS})\boldsymbol{\nu}_i\right]\right\} \quad (3.118)$$

$$\approx D\left(\frac{\partial \mu^{SS}}{\partial \eta^{SS}}\right)\boldsymbol{\nu}_i\boldsymbol{\Sigma}_\nu\boldsymbol{\nu}_i^T D\left(\frac{\partial \mu^{SS}}{\partial \eta^{SS}}\right) + \phi V(\mu^{SS}) = \tilde{V}(\mathbf{y}_i) \quad (3.119)$$

3.3.3 Applications of the SS-GEE

This section illustrates a model for linear regression for the purpose of illustrating SS-GEE. The majority of software packages offer several competing methods for fitting this model, and we wish to highlight the equivalence of the PA-GEE and SS-GEE for this particular case.

For the case of a single Gaussian distributed random effect, we can derive the exact solution for the variance as

$$V^{SS}(y_{it}) = \int \left[(\mu_{is} + \nu_i) - \mu_{is}\right]\left[(\mu_{it} + \nu_i) - \mu_{it}\right]df(\nu_i) +$$

$$\phi I(s=t)\int df(\nu_i) \quad (3.120)$$

$$= JJ^T\sigma_\nu^2 + D(\phi) \quad (3.121)$$

such that the panel variance is given by

$$V = \begin{bmatrix} \phi + \sigma_\nu^2 & \sigma_\nu^2 & \sigma_\nu^2 & \cdots & \sigma_\nu^2 \\ \sigma_\nu^2 & \phi + \sigma_\nu^2 & \sigma_\nu^2 & \cdots & \sigma_\nu^2 \\ \sigma_\nu^2 & \sigma_\nu^2 & \phi + \sigma_\nu^2 & \cdots & \sigma_\nu^2 \\ \vdots & \vdots & \vdots & \ddots & \vdots \\ \sigma_\nu^2 & \sigma_\nu^2 & \sigma_\nu^2 & \cdots & \phi + \sigma_\nu^2 \end{bmatrix} \quad (3.122)$$

This is, in fact, the same hypothesized structure as the exchangeable correlation PA-GEE that we observed for population averaged models. The two models are equivalent since the link functions for the PA and SS models are the same. However, they may differ in calculation depending on the method used to estimate the panel variance components.

Since the two models are the same, there is no compelling reason to calculate this particular SS model—software already exists for the equivalent PA model. In addition, the software packages referred to in this text also include the means to estimate an equivalent FIML model (Gaussian distributed random effects linear regression). Fitting the PA model allows us an interpretation under either the PA or the SS model assumptions.

We can still derive the appropriate panel level variance component if we want to fit a linear regression model with more than one random effect. Such a model is equivalent to a mixed model that is supported by other software (see PROC MIXED in SAS or the user-contributed gllamm command in Stata). Under these conditions, it is again not compelling to go through the trouble of programming the SS-GEE since software exists for the equivalent models.

On the other hand, we can consider programming the resulting estimator if we wish to fit a log-linear regression model (using the log link rather than the identity link). We already saw that the random effects induce an offset on the link function that differs from a similar population averaged model.

Another illustration is the SS-GEE Poisson model for the Progabide data. We saw earlier that the maximum likelihood Gaussian random effects model results in

```
Random-effects poisson            Number of obs       =      295
Group variable (i) : id           Number of groups    =       59

Random effects u_i ~ Gaussian     Obs per group: min  =        5
                                                 avg  =      5.0
                                                 max  =        5

                                  LR chi2(3)          =    33.40
Log likelihood = -1017.4249       Prob > chi2         =   0.0000
```

THE SS-GEE FOR GLMS

```
    seizures |      Coef.   Std. Err.      z    P>|z|     [95% Conf. Interval]
-------------+----------------------------------------------------------------
        time |    .111836    .0468768    2.39   0.017     .0199591    .2037129
    progabide |  -.5396357    .0545001   -9.90   0.000    -.6464539   -.4328174
   timeXprog |  -.1047258    .0650304   -1.61   0.107     -.232183    .0227314
       _cons |   1.124231    .0430617   26.11   0.000     1.039831    1.20863
    lnPeriod |   (offset)
-------------+----------------------------------------------------------------
    /lnsig2u |  -.8970602    .0495843  -18.09   0.000    -.9942437   -.7998767
-------------+----------------------------------------------------------------
     sigma_u |   .6385661    .0158314                     .6082788    .6703614
         rho |    .289655    .0102022                     .2700747    .3100519
------------------------------------------------------------------------------
Likelihood ratio test of rho=0:    chibar2(01) =   2602.16 Prob>=chibar2 = 0.000
```

The above results are calculated using a straightforward Gauss–Hermite quadrature approximation of the likelihood, gradient, and Hessian. This application offers another opportunity for us to illustrate the sensitivity of this approximation for rough functions. Using an adaptive quadrature approximation, we obtain the following results

```
    seizures |      Coef.   Std. Err.      z    P>|z|     [95% Conf. Interval]
-------------+----------------------------------------------------------------
        time |   .1118361    .0468766    2.39   0.017     .0199597    .2037125
    progabide |  .0051622    .0530336    0.10   0.922    -.0987817    .1091062
   timeXprog |   -.104726    .0650299   -1.61   0.107    -.2321823    .0227303
       _cons |   1.069857    .0480689   22.26   0.000     .9756434    1.16407
    lnPeriod |   (offset)
------------------------------------------------------------------------------

Variances and covariances of random effects
------------------------------------------------------------------------------
***level 2 (id)

    var(1):  .2970534  (.01543218)
------------------------------------------------------------------------------
```

The most striking difference in the two maximum likelihood models is the change in sign for the progabide variable. This difference is a reflection of the sensitivity of the straightforward Gauss–Hermite quadrature approximation used in the first model.

Finally, we can fit a SS-GEE Poisson model to the data for comparison

```
    seizures |      Coef.     Std. Err.
-------------+---------------------------
        time |  .11843141    .04901529
    progabide| .02746961    .00899939
   timeXprog |-.10386892    .07324486
       _cons |  1.3499396    .00669060
    lnPeriod |   (offset)
-------------+---------------------------
    sigma^2_v|  .15731536
------------------------------------------
```

The results differ from both of the approximated maximum likelihood methods given before. In fact, the model is not a good choice for the data regardless of the calculation since there is evidence of overdispersion that we have not addressed. One of the points we have emphasized in this section is that even though we can program a SS-GEE model, it is often unnecessary. The SS-GEE model includes the assumption that the estimating equation for the variance parameters is orthogonal to the estimating equation for the regression parameters. Most software packages having maximum likelihood are uncorrelated.

We can fit a maximum likelihood gamma distributed random effects model for this particular model. The gamma distributed random effects model has an estimated covariance matrix for this data that is nearly zero for the covariance of the random effects parameter and the regression parameters, which is enforced by the orthogonality assumption imposed by the SS-GEE model. The results of this model more closely agree with the SS-GEE results than did the Gaussian distributed random effects models. To be clear, the gamma distributed random effects model does not impose an assumption of orthogonality. The small estimates of covariance for the estimating equations of the regression and variance parameters for this particular dataset are not a general result.

```
Random-effects Poisson                          Number of obs      =       295
Group variable (i) : id                         Number of groups   =        59

Random effects u_i ~ Gamma                      Obs per group: min =         5
                                                               avg =       5.0
                                                               max =         5

                                                Wald chi2(3)       =      5.73
Log likelihood  = -1017.3826                    Prob > chi2        =    0.1253

------------------------------------------------------------------------------
    seizures |     Coef.   Std. Err.      z    P>|z|     [95% Conf. Interval]
-------------+----------------------------------------------------------------
        time |   .111836    .0468768     2.39   0.017     .0199591    .2037129
    progabide|  .0275345    .2108952     0.13   0.896    -.3858125    .4408815
    timeXprog| -.1047258    .0650304    -1.61   0.107     -.232183    .0227314
       _cons |  1.347609    .1529187     8.81   0.000     1.047894    1.647324
    lnPeriod |  (offset)
-------------+----------------------------------------------------------------
    /lnalpha | -.474377     .1731544                     -.8137534   -.1350007
-------------+----------------------------------------------------------------
       alpha |  .6222726    .1077492                      .4431915    .8737153
------------------------------------------------------------------------------
Likelihood ratio test of alpha=0: chibar2(01) =  2602.24 Prob>=chibar2 = 0.000
```

3.3.4 Estimating the SS-GEE model

The complete SS-GEE is given by

$$\Psi(\beta, \alpha) = \left(\sum_{i=1}^{n} \mathbf{x}_{ji}^{\mathrm{T}} \mathrm{D} \left(\frac{\partial \mu_i}{\partial \eta} \right) (\mathrm{V}(\mu_i))^{-1} \left(\frac{\mathbf{y}_i - \mu_i}{a(\phi)} \right) \right) \quad (3.123)$$

$$\mu_i = \int f(\nu_i) g^{-1}(\mathbf{x}_{it}\beta^{\mathrm{SS}} + \nu_i) d\nu_i \quad (3.124)$$

$$\widetilde{\mathrm{V}}(\mu_i) \approx \mathrm{D}\left(\frac{\partial \mu^{\mathrm{SS}}}{\partial \eta^{\mathrm{SS}}}\right) \nu_i \mathbf{\Sigma}_\nu(\alpha) \nu_i^{\mathrm{T}} \mathrm{D}\left(\frac{\partial \mu^{\mathrm{SS}}}{\partial \eta^{\mathrm{SS}}}\right) + \phi \mathrm{V}(\mu^{\mathrm{SS}}) \quad (3.125)$$

$$\mathbf{\Sigma}_\nu(\alpha) = \text{Parameterized variance matrix} \quad (3.126)$$

Specification of the second estimating equation $\Psi(\alpha, \beta)$ for calculating the components of $\mathbf{\Sigma}_\nu(\alpha)$ for specific subject-specific models is left to the reader.

At each step in the usual GLM algorithm, we first estimate the variance components $\mathbf{V}(\mu)$, and then use that result to estimate β. At the initial step we can assume a diagonal matrix for the variances. Calculation of the components of the variance matrix involves estimating the dispersion parameter.

As in the case of the PA-GEE models, we can calculate the sandwich estimate instead of the naive variance estimate. Doing so further protects us from misspecification of the within-panel correlation assumption implied by the variance component.

Recall that the modified sandwich estimate of variance is robust to *any* form of within panel correlation. Thus, we gain efficiency in our estimated β if we have the correct form of the within panel covariances, and we are protected from misspecification if we are wrong.

The modified sandwich estimate of variance for the complete estimating equation is found using the same approach as was used for PA-GEE models

$$\mathrm{V}_{\mathrm{MS}} = \mathbf{A}^{-1}\mathbf{B}\mathbf{A}^{-\mathrm{T}} \quad (3.127)$$

$$\mathbf{A} = \begin{bmatrix} -\frac{\partial \Psi_\beta}{\partial \beta} & -\frac{\partial \Psi_\beta}{\partial \alpha} \\ -\frac{\partial \Psi_\alpha}{\partial \beta} & -\frac{\partial \Psi_\alpha}{\partial \alpha} \end{bmatrix}^{-1} \quad (3.128)$$

$$\mathbf{B} = \sum_{i=1}^{n} \left(\sum_{t=1}^{n_i} \begin{pmatrix} \Psi_{\beta it} \\ \Psi_{\alpha it} \end{pmatrix} \right) \left(\sum_{t=1}^{n_i} \begin{pmatrix} \Psi_{\beta it} \\ \Psi_{\alpha it} \end{pmatrix} \right)^{\mathrm{T}} \quad (3.129)$$

Again, since we assume that the two estimating equations are orthogonal, we can write

$$\mathbf{A} = \begin{bmatrix} -\frac{\partial \Psi_\beta}{\partial \beta} & 0 \\ 0 & -\frac{\partial \Psi_\alpha}{\partial \alpha} \end{bmatrix}^{-1} = \begin{bmatrix} \left(-\frac{\partial \Psi_\beta}{\partial \beta}\right)^{-1} & 0 \\ 0 & \left(-\frac{\partial \Psi_\alpha}{\partial \alpha}\right)^{-1} \end{bmatrix} \quad (3.130)$$

This assumption of orthogonality leads to a block diagonal modified sandwich estimate of variance where the upper left entry is given by

$$\left(-\frac{\partial \Psi_\beta}{\partial \beta}\right) \left[\sum_{i=1}^{n} \left(\sum_{t=1}^{n_i} \Psi_{\beta it}\right) \left(\sum_{t=1}^{n_i} \Psi_{\beta it}\right)^{\mathrm{T}}\right] \left(-\frac{\partial \Psi_\beta}{\partial \beta}\right) \quad (3.131)$$

The lower right entry is given by

$$\left(-\frac{\partial \Psi_\alpha}{\partial \alpha}\right) \left[\sum_{i=1}^{n} \left(\sum_{t=1}^{n_i} \Psi_{\alpha it}\right) \left(\sum_{t=1}^{n_i} \Psi_{\alpha it}\right)^{\mathrm{T}}\right] \left(-\frac{\partial \Psi_\alpha}{\partial \alpha}\right) \quad (3.132)$$

Because we are interested only in the regression parameters, and since the matrix is block diagonal, there is no need to calculate variances for the ancillary parameters. The modified sandwich estimate of variance for the regression parameters is the upper $p \times p$ part of V_{MS} given by equation 3.131 and the modified sandwich estimate of variance for the association parameters is the lower $q \times q$ part of V_{MS} given in equation 3.132. The variance of the association parameters is not calculated in the approach given in section 3.2, though the formula is valid. The variance of the association parameters is calculated in the ALR approach (section 3.2.5) as well as other GEE models.

We advise that all users specify the modified sandwich estimate of variance with this model. We are free to use either the expected Hessian or the observed Hessian as we did for the PA-GEE model.

3.3.5 Summary

The SS-GEE is not implemented as often as the PA-GEE model. The main reason is that alternatives such as maximum likelihood are typically available. Maximum likelihood methods, when available, estimate the same population parameter. All of the GEE models that assume orthogonality of the estimating equations for the regression parameters as well as the association parameters are called "GEE of order 1" or "GEE1."

We should also emphasize that the focus of the PA-GEE model already covered was the introduction of structured correlation. Since the marginal model is introduced directly into the estimation of the PA-GEE model, and we restrict attention to the within-panel correlation, most of the resulting variance structures implied by the correlation can not even approximately be generated from a random-effects model. If the variance structures are a focus of the analysis and we believe a mixed model explains the data, we should focus attention on a SS-GEE (or an equivalent likelihood-based model) over a PA-GEE model.

3.4 The GEE2 for GLMs

Our discussion of the PA-GEE for GLMs included two estimating equations; one is for estimating β and the other for estimating α. Since we were not interested in the correlation parameters, and since we assumed that the two

THE GEE2 FOR GLMS

coefficient vectors were orthogonal, we focused only on the estimating equation for β treating α as ancillary.

Formally, our overall GEE2 for GLMs addresses both of the parameter vectors and their associated estimating equations. In this case there is no assumption that the two estimating equations are orthogonal. The GEE2 may be written

$$\sum_{i=1}^{n} \begin{pmatrix} \frac{\partial \boldsymbol{\mu}_i}{\partial \boldsymbol{\beta}} & \frac{\partial \boldsymbol{\mu}_i}{\partial \boldsymbol{\alpha}} \\ \frac{\partial \boldsymbol{\sigma}_i}{\partial \boldsymbol{\beta}} & \frac{\partial \boldsymbol{\sigma}_i}{\partial \boldsymbol{\alpha}} \end{pmatrix}^{\mathrm{T}} \begin{pmatrix} V(\mathbf{y}_i, \mathbf{y}_i) & V(\mathbf{y}_i, \mathbf{s}_i) \\ V(\mathbf{s}_i, \mathbf{y}_i) & V(\mathbf{s}_i, \mathbf{s}_i) \end{pmatrix}^{-1} \begin{pmatrix} \mathbf{y}_i - \boldsymbol{\mu}_i \\ \mathbf{s}_i - \boldsymbol{\sigma}_i \end{pmatrix} = [\mathbf{0}]$$

(3.133)

What GEE1 does differently from GEE2 is to assume that the first two terms in the estimating equation are block diagonal (assume zero matrices in the off diagonal positions). It is therefore clear that GEE2 is a generalization of GEE1.

For GEE2, we not only have to provide a working correlation matrix for the regression parameters, but we must also provide a working covariance matrix for the correlation parameters. In other words, instead of making assumptions on the first 2 moments, we make assumptions on the first 4 moments. It is much more difficult to picture these assumptions and understand the constraints that result. It can be difficult to interpret the mean vector that includes a dependence on the association parameters in the GEE2 specification. In most applications the mean is only defined by the regression parameters; but the assumption that $\partial \boldsymbol{\mu}_i / \partial \boldsymbol{\alpha} \neq \mathbf{0}$ implies that the correlation is a function of the regression parameters $\boldsymbol{\beta}$.

The question is how to define the estimating equations to model the association of the covariance in terms of both the regression parameters and the association parameters. The GEE2 specification is not often used because of the fact that to obtain a consistent estimator of the regression parameters, we must correctly specify the link function as well as the covariance function. If we are willing to assume that a block diagonal parameterization is correct, the procedure yields consistent estimation of the regression parameters – even if the association is incorrectly specified. Certain models have been proposed, with appropriate proofs, dealing with the consistency and distribution of the results. Results for the asymptotic covariance matrix and consistency of the estimators are given in Zhao and Prentice (1990), Prentice and Zhao (1991), and Gourieroux and Monfort (1993).

A sandwich estimate of variance is constructed in the usual way and illustrates that the $\mathbf{A}^{-1}$ matrix in the definition of the sandwich estimate of variance $\mathbf{A}^{-1}\mathbf{B}\mathbf{A}^{-\mathrm{T}}$ is not necessarily symmetric—it is block diagonal where each block is symmetric. As pointed out earlier, while the sandwich estimate of variance is robust to misspecification of the correlation structure for the PA-GEE model, it does not have this property for the GEE2 models. The reason is that the GEE2 models do not assume the orthogonality of the two estimating equations. Hence, in this situation the hypothesized correlation structure enters into the calculation of the sandwich estimate of variance.

For ease of presentation, note that the overall estimating equation may be written for $\Theta = (\beta, \alpha)$ as

$$\Psi(\Theta) = (\Psi_\beta(\beta, \alpha) \; \Psi_\alpha(\beta, \alpha)) \tag{3.134}$$

such that the sandwich estimate of variance is given by

$$V_{MS} = \mathbf{A}^{-1} \mathbf{B} \mathbf{A}^{-T} \tag{3.135}$$

where

$$\mathbf{A}^{-1} = \begin{bmatrix} -\dfrac{\partial \Psi_\beta}{\partial \beta} & -\dfrac{\partial \Psi_\beta}{\partial \alpha} \\ -\dfrac{\partial \Psi_\alpha}{\partial \beta} & -\dfrac{\partial \Psi_\alpha}{\partial \alpha} \end{bmatrix} \tag{3.136}$$

and

$$\mathbf{B} = \sum_{i=1}^{n} \left(\sum_{t=1}^{n_i} \Psi_{it} \right) \left(\sum_{t=1}^{n_i} \Psi_{it} \right)^{T} \tag{3.137}$$

3.5 GEEs for extensions of GLMs

There have been notable extensions of classes of GEE models to multinomial data. Multinomial data are classified as data where the response variable takes on one of several distinct outcomes—the complete set of outcomes may or may not have a natural order.

3.5.1 Generalized logistic regression

The generalized logistic regression model assumes that the response counts of each covariate pattern have a multinomial distribution, where the multinomial counts for different covariate patterns are independent. Due to these assumptions, this type of model is usually called the multinomial logit model.

Upon assuming that one of the outcomes is the reference outcome, the model simultaneously fits a logistic regression model comparing each of the other outcomes to the reference. As such, with k possible outcomes, there are $(k-1)$ logistic regression vectors. For this model, the exponentiated coefficients are not always called odds ratios since they denote the odds of being in category j instead of the reference category; they are sometimes called relative risk ratios.

The SUDAAN package is the only one of the four packages used in this text that has support for fitting this model. This package may be used in standalone mode or as a callable `PROC` from the SAS package. Our examples use the SAS callable method. Shah, Barnwell, and Bieler (1997) can be referenced for documentation on using the software as well as statistical documentation on the methods implemented. We should mention that this particular package emphasizes the analysis of complex survey data (not covered in this text). As discussed previously, SUDAAN software can be used without specification of the sampling frame.

In the following output we use a constructed dataset with 50 panels of size 10. Category 3 of the response variable is used as the reference.

```
Independence parameters have converged in 5 iterations

Step 1 parameters have converged in 8 iterations.

Number of observations read      :   500    Weighted count:    500
Observations used in the analysis :   500    Weighted count:    500
Denominator degrees of freedom   :    49

Maximum number of estimable parameters for the model is  6

File GEE contains    50 Clusters
  50 clusters were used to fit the model
Maximum cluster size is  10 records
Minimum cluster size is  10 records

Sample and Population Counts for Response Variable Y
  1:  Sample Count     105    Population Count      105
  2:  Sample Count     260    Population Count      260
  3:  Sample Count     135    Population Count      135

Variance Estimation Method: Taylor Series (WR)
SE Method: Robust (Binder, 1983)
Working Correlations: Exchangeable
Link Function: Generalized Logit
Response variable Y: Y
-----------------------------------------------------------------------
|                |                 |                                   |
| Y (log-odds)   |                 | Independent Variables and Effects |
|                |                 | Intercept  | X1        | X2       |
-----------------------------------------------------------------------
|                |                 |            |           |          |
| 1 vs 3         | Beta Coeff.     |    0.36    |   -1.42   |   -1.27  |
|                | SE Beta         |    0.20    |    0.39   |    0.23  |
|                | T-Test B=0      |    1.77    |   -3.66   |   -5.57  |
|                | P-value T-Test  |            |           |          |
|                |   B=0           |    0.0823  |    0.0006 |    0.0000|
-----------------------------------------------------------------------
|                |                 |            |           |          |
| 2 vs 3         | Beta Coeff.     |    1.32    |   -1.11   |   -0.91  |
|                | SE Beta         |    0.18    |    0.26   |    0.14  |
|                | T-Test B=0      |    7.39    |   -4.24   |   -6.69  |
|                | P-value T-Test  |            |           |          |
|                |   B=0           |    0.0000  |    0.0001 |    0.0000|
-----------------------------------------------------------------------

Correlation Matrix

-------------------------------------------
Y                 Y
                       1          2
-------------------------------------------
1                 0.0190
2                -0.0123     0.0272
-------------------------------------------
```

The output includes the estimated correlation parameters for the exchangeable model. Note that there are three correlation parameters listed instead of the single parameter one might expect. This model is a simultaneous estimation of logistic regression models comparing outcomes to the reference category, such that we allow a different common correlation for the comparisons.

3.5.2 Cumulative logistic regression

An alternative approach assumes that the outcomes do have a natural ordering. In this approach, the cumulative logits are used to define the model. In addition to the $(k-1)$ sets of regression parameters, there are also $(k-1)$ cut points estimated for the k possible outcomes. The exponentiated regression coefficients are interpreted as the odds of being in category k or lower. The response curves all have the same shape; the effects of the covariates are assumed to be the same for each of the cut points.

There is a danger of confusion with this model since different terminology is used in competing software manuals. SAS software labels this the cumulative logistic regression model while Stata refers to it as the ordered logistic regression model. Stata offers no support for ordered models using the PA-GEE software. SAS will allow users to specify the model, but does not support any correlation structure except independence. Stata offers pooled ordered logistic and probit regression models via the ologit and oprobit commands. These two Stata commands also have the option to allow calculation of the sandwich estimate of variance. In addition to these estimators, SAS also supports the cumulative complementary log-log model; again, only with the independent correlation structure specification.

To emphasize, we assume that there are k possible ordered outcomes. In addition to the coefficients of the model, cut points are also estimated to divide the range of the outcome into k categories. Some software packages will directly list the cutpoints and others will list "intercepts"; these intercepts are equal to the negatives of the cutpoints.

SAS Institute, Inc. (2000) includes a simple example to illustrate the cumulative logistic regression model. We have data on three different brands of ice cream. Included are the number of people (count) who ranked the ice cream (brand) on a Likert scale (taste) from 1 to 5; 1=very bad, 2=bad, 3=average, 4=good, 5=very good. The outcome variable is the qualitative taste of the ice cream for each brand. The count variable is a replication style weight indicating the number of individuals assigning the associated taste category to that associated ice cream brand. Replication weights are sometimes called data compression weights or frequency weights.

The data are:

count	brand	taste
70	1	5
71	1	4
151	1	3
30	1	2
46	1	1
20	2	5
36	2	4
130	2	3
74	2	2
70	2	1
50	3	5
55	3	4
140	3	3
52	3	2
50	3	1

The cumulative logistic regression results obtained from SAS are displayed as:

Criteria For Assessing Goodness Of Fit

Criterion	DF	Value	Value/DF
Log Likelihood		-1564.2269	

Analysis Of Parameter Estimates

Parameter	DF	Estimate	Standard Error	Wald 95% Confidence Limits		Chi-Square	Pr > ChiSq
Intercept1	1	-1.4936	0.1585	-1.8041	-1.1830	88.85	<.0001
Intercept2	1	-0.5231	0.1485	-0.8142	-0.2320	12.40	0.0004
Intercept3	1	1.1981	0.1533	0.8977	1.4985	61.10	<.0001
Intercept4	1	2.0595	0.1631	1.7399	2.3792	159.47	<.0001
brand	1	-0.1932	0.0681	-0.3266	-0.0597	8.05	0.0045
Scale	0	1.0000	0.0000	1.0000	1.0000		

The ordered logistic regression results obtained from Stata are:

```
Ordered logit estimates                         Number of obs   =      1045
                                                LR chi2(1)      =      8.08
                                                Prob > chi2     =    0.0045
Log likelihood = -1564.2269                     Pseudo R2       =    0.0026

------------------------------------------------------------------------------
       taste |      Coef.   Std. Err.       z    P>|z|     [95% Conf. Interval]
-------------+----------------------------------------------------------------
       brand |  -.1931679   .0680729    -2.84    0.005    -.3265883   -.0597475
-------------+----------------------------------------------------------------
       _cut1 |  -2.059537   .1630906           (Ancillary parameters)
       _cut2 |  -1.198094   .1532729
       _cut3 |   .5230929   .1485218
       _cut4 |   1.493574   .1584554
------------------------------------------------------------------------------
```

Corresponding likelihood models can be developed for each of the models within the standard GLM framework. For example, a random-effects ordered model can be fit assuming a Gaussian distribution for the random effects. Typically, software implementations of this model will use the Gauss–Hermite quadrature approximation that we have discussed. For the ice cream data analyzed earlier, we now assume that there is a random effect associated with each brand.

The results of fitting a Gaussian distributed random-effects ordered probit regression model to the ice cream data are:

```
Random-effects ordered probit estimates          Number of obs   =       1045
Log likelihood = -1548.3929

------------------------------------------------------------------------------
       taste |      Coef.   Std. Err.       z    P>|z|     [95% Conf. Interval]
-------------+----------------------------------------------------------------
_cut1        |
       _cons |  -.9368743    .0479922   -19.52   0.000    -1.030937   -.8428114
-------------+----------------------------------------------------------------
_cut2        |
       _cons |  -.4278815    .0423327   -10.11   0.000    -.5108521   -.3449109
-------------+----------------------------------------------------------------
_cut3        |
       _cons |   .6603478    .0442729    14.92   0.000     .5735745    .7471211
-------------+----------------------------------------------------------------
_cut4        |
       _cons |   1.226776    .0524694    23.38   0.000     1.123938    1.329615
-------------+----------------------------------------------------------------
rho          |
       _cons |   .2286464    .0518036     4.41   0.000     .1271133    .3301795
------------------------------------------------------------------------------
```

Both the ordered logit and random-effects ordered probit model are likelihood-based so that we can choose between the models using criteria such as the Akaike information criterion (AIC) or the Bayesian information criterion (BIC). Using the AIC, we prefer the ordered logit model over the random-effects ordered probit model.

3.6 Further developments and applications

Research continues in the several areas of generalized estimating equations. We present an introduction to some of the recently proposed applications and theory in the following sections.

3.6.1 The PA-GEE for GLMs with measurement error

Here we describe a method for generating a valid variance estimate for the case of a PA-GEE with instrumental variables for measurement error. Obtaining point estimates from the model is relatively straightforward since we simply replace the endogenous regressors with the predicted values from OLS regressions. However, a valid variance estimate of the PA-GEE regression parameters must take into account the error associated with the instrumental variables regressions.

In this section, we introduce a notation varying from the usual notation associated with measurement error models. The usual notation involves naming individual matrices: $\mathbf{Z}$ for covariates measured without error, $\mathbf{W}$ for covariates measured with error, $\mathbf{S}$ for the instruments of $\mathbf{W}$, and $\mathbf{R}$ for the augmented matrix of exogenous variables $[\mathbf{Z}\ \mathbf{S}]$. In order to avoid confusion with the measurement error notation and the usual notation associated with GLMs and PA-GEEs (the $\mathbf{W}$ weight matrix in the IRLS algorithm and the $\mathbf{R}$ working correlation matrix of the Liang and Zeger PA-GEE), we demote the measurement error matrix notational conventions to subscripts of the $\mathbf{X}$ matrix in the PA-GEE notation.

We begin with an $n \times p$ matrix of covariates measured without error given by the augmented matrix $\mathbf{X} = (\mathbf{X}_1\ \mathbf{X}_2)$ and consider the case for which $\mathbf{X}_1 = \mathbf{X}_Z$, where $\mathbf{X}_2$ is unobserved, and $\mathbf{X}_W = \mathbf{X}_2$ plus measurement error. $\mathbf{X}_Z$ is a $n \times p_z$ matrix of covariates measured without error (possibly including a constant), and $\mathbf{X}_W$ is a $n \times p_x$ ($p_z + p_x = p$) matrix of covariates with classical measurement error that estimates $\mathbf{X}_2$. We wish to utilize an $n \times p_s$ (where $p_s \geq p_x$) matrix of instruments $\mathbf{X}_S$ for $\mathbf{X}_W$.

Greene (2000) discusses instrumental variables and provides a clear presentation to supplement the following concise description. The method of instrumental variables assumes that some subset $\mathbf{X}_W$ of the independent variables is correlated with the error term in the model. In addition, we have a matrix $\mathbf{X}_S$ of independent variables which are correlated with $\mathbf{X}_W$. $\mathbf{X}$, $\mathbf{Y}$, and $\mathbf{X}_W$ are uncorrelated. Using these relationships, we can construct an approximately consistent estimator that may be succinctly described. We estimate a regression for each of the independent variables (each column) of $\mathbf{X}_W$ on the instruments and the independent variables not correlated with the error term ($\mathbf{X}_Z\ \mathbf{X}_S$). Predicted values are then obtained from each regression and substituted for the associated column of $\mathbf{X}_W$ in the analysis of the PA-GEE of interest. This construction provides an approximately consistent estimator of the coefficients in the PA-GEE (it is consistent in the linear case).

If we have access to the complete matrix of covariates measured without error (if we know $\mathbf{X}_2$ instead of just $\mathbf{X}_W$), we denote the linear predictor $\eta = \sum_{j=1}^{p} [\mathbf{X}_1\ \mathbf{X}_2]_j \beta_j$, and the associated derivative as $\partial \eta / \partial \beta_j = [\mathbf{X}_1\ \mathbf{X}_2]_j$. The estimating equation for β is then $\sum_{i=1}^{n} (y_i - \mu_i)/V(\mu_i)(\partial \mu / \partial \eta)_i [\mathbf{X}_1\ \mathbf{X}_2]_{ji}$.

However, since we do not know $\mathbf{X}_2$, we use $\mathbf{X}_R = (\mathbf{X}_Z\ \mathbf{X}_S)$ to denote the augmented matrix of exogenous variables, which combines the covariates measured without error and the instruments. We regress each of the p_x components (each of the p_x columns) of $\mathbf{X}_W$ on $\mathbf{X}_R$ to obtain an estimated $(p_z + p_s) \times 1$ coefficient vector γ_j for $j = 1, \ldots, p_x$.

The naive variance estimate is not valid when instrumental variables are used. For this case, we must rely on other asymptotic estimates of the variance that take into account the instrumental variables regressions. One such estimate presented here is the sandwich estimate of variance.

While we are ultimately interested in β, we must consider all of the (nonancillary) parameters from the instrumental variables regressions in forming the associated variance matrix. It is not reasonable to assume that these two coefficient vectors are orthogonal.

To include this approach in a PA-GEE for instrumental variable GLMs, we must first write the full estimating equation for $\Theta = (\beta, \gamma, \alpha)$ as

$$\Psi(\Theta) = (\Psi_\beta(\beta,\gamma,\alpha), \Psi_\gamma(\beta,\gamma,\alpha) \Psi_\alpha(\beta,\gamma,\alpha)) \quad (3.138)$$

where the individual estimating equations, including the matrix sizes, are given by

$$\Psi_\beta = \left(\sum_{i=1}^n [\mathbf{X}_Z\ \mathbf{X}_R\widehat{\gamma}]_{ji} D\left(\frac{\partial \mu_i}{\partial \eta}\right) V(\mu_i)^{-1}\left(\frac{\mathbf{y}_i - \mu_i}{a(\phi)}\right)\right)_{p\times 1}^{j=1,\ldots,p} \quad (3.139)$$

$$\Psi_\gamma = \begin{pmatrix} \sum_{i=1}^n \left[(\mathbf{X}_{W1} - \mathbf{X}_R\gamma_{1j})\mathbf{X}_{Rji}\right]_{(p_z+p_s)\times 1}^{j=1,\ldots,(p_z+p_s)} \\ \sum_{i=1}^n \left[(\mathbf{X}_{W2} - \mathbf{X}_R\gamma_{2j})\mathbf{X}_{Rji}\right]_{(p_z+p_s)\times 1}^{j=1,\ldots,(p_z+p_s)} \\ \vdots \\ \sum_{i=1}^n \left[(\mathbf{X}_{Wp_x} - \mathbf{X}_R\gamma_{p_xj})\mathbf{X}_{Rji}\right]_{(p_z+p_s)\times 1}^{j=1,\ldots,(p_z+p_s)} \end{pmatrix}_{(p_x(p_z+p_s))\times 1} \quad (3.140)$$

$$\Psi_\alpha = \left(\sum_{i=1}^n \left(\frac{\partial \xi_i}{\partial \alpha}\right)^T \mathbf{H}_i^{-1}(\mathbf{W}_i - \xi_i)\right)_{q\times 1} \quad (3.141)$$

Estimation is performed in two stages. First we run the OLS regressions for the instrumental variables. Predicted values are obtained from the OLS regressions and used as proxies for the appropriate variables in the generalized linear model of interest. Subsequent estimation is the same as in PA-GEE. An estimate of β is obtained followed by estimates of the ancillary parameters α and ϕ. We alternate estimation between the coefficient vector and the ancillary parameters until subsequent estimates of β are within some convergence criterion.

The upper $p \times p$ submatrix of $\mathbf{A}$ is a naive variance estimate that is not valid for the GLM regression parameters β, since it assumes that the fitted values from the OLS instrumental variables' regressions are true (without error). In addition, that matrix assumes that β and γ are orthogonal. In most cases, the instruments that we use in the OLS are a subset of the GLM covariates, so that this assumption is untenable. A valid variance estimate may be obtained using the modified sandwich estimate of variance given in this section.

Our goal is to calculate the modified sandwich estimate of variance given by $\mathbf{V}_{\mathrm{MS}} = \mathbf{A}^{-1}\,\mathbf{B}\,\mathbf{A}^{-\mathrm{T}}$. We form the variance matrix, $\mathbf{A}$, for Θ by obtaining the necessary derivatives where both $\boldsymbol{\beta}$ and $\boldsymbol{\gamma}$ are each assumed to be orthogonal to $\boldsymbol{\alpha}$; but $\boldsymbol{\beta}$ and $\boldsymbol{\gamma}$ are not assumed to be orthogonal to each other.

$$\mathbf{A}^{-1} = \begin{bmatrix} -\dfrac{\partial \Psi_\beta}{\partial \beta}_{p\times p} & -\dfrac{\partial \Psi_\beta}{\partial \gamma}_{p\times (p_x(p_z+p_s))} & 0 \\ 0 & -\dfrac{\partial \Psi_\gamma}{\partial \gamma}_{(p_x(p_z+p_s))\times (p_x(p_z+p_s))} & 0 \\ 0 & 0 & -\dfrac{\partial \Psi_\alpha}{\partial \alpha}_{q\times q} \end{bmatrix}^{-1} \quad (3.142)$$

The $(2,1)$ submatrix of $\mathbf{A}$ is zero since $\boldsymbol{\beta}$ does not enter the Ψ_γ estimating equation for the OLS regressions; but the $(1,2)$ submatrix is not zero since $\boldsymbol{\gamma}$ enters the Ψ_β estimating equation through the predicted values of the OLS regressions. The other zero submatrices are the result of assumptions of orthogonality, though one could develop a GEE2 without this assumption.

The middle of the modified sandwich estimate of variance is given by

$$\mathbf{B} = \sum_{i=1}^{n} \left(\sum_{t=1}^{n_i} \begin{pmatrix} \Psi_{\beta it} \\ \Psi_{\gamma it} \\ \Psi_{\alpha it} \end{pmatrix} \right) \left(\sum_{t=1}^{n_i} \begin{pmatrix} \Psi_{\beta it} \\ \Psi_{\gamma it} \\ \Psi_{\alpha it} \end{pmatrix} \right)^{\mathrm{T}} \quad (3.143)$$

and the sandwich estimate of variance for $\boldsymbol{\beta}$ is the upper $p \times p$ matrix of $\mathbf{V}_{\mathrm{MS}}$.

The derivation of a valid variance estimate for PA-GEE analysis of panel data is an application of the general method of forming sandwich estimates of variance. Obtaining estimates for the regression parameter vector $\boldsymbol{\beta}$ is relatively straightforward using most statistical packages; but obtaining the sandwich estimate of variance requires some work on the part of the user. We can not simply replace the covariates with the OLS regression predicted values and fit a PA-GEE model. The standard errors (naive or modified sandwich) are not correct and you must follow the derivation given above to construct a valid estimate.

Sandwich estimates of variance are formed using $\mathbf{V}_{\mathrm{MS}} = \mathbf{A}^{-1}\mathbf{B}\mathbf{A}^{-\mathrm{T}}$. $\mathbf{A}$ is symmetric for many applications. However, as Binder (1992) points out, the bread of the sandwich estimate of variance is not, in general, symmetric. The asymmetry in the case of GLMs for longitudinal data with instrumental variables due to the augmented matrices of cross derivatives is such an example.

In constructing sandwich estimates of variance, the bread of the sandwich (the $\mathbf{A}$ matrix) is the matrix of second derivatives of the complete estimating equation. The middle of the sandwich estimate of variance (the $\mathbf{B}$ matrix) is the variance of the complete estimating equation. In many cases, the estimating equation involves independent observations; and the $\mathbf{B}$ matrix may be formed as the sum (over observations) of the outer product of the estimating equation $\sum_i \Psi_i \Psi_i^T$.

To highlight this application, we present a contrived example with a very small dataset so that the construction of the relevant matrices is clear. We also point out that the sandwich estimate of variance to be calculated for the regression coefficients does not depend on the hypothesized correlation structure. The regression coefficients are certainly affected, but the sandwich estimate of variance is a post-estimation adjustment. To see this, organize the $\mathbf{A}$ and $\mathbf{B}$ matrices that go into the calculation as:

$$\mathbf{A}^{-1} = \left[\begin{pmatrix} -\frac{\partial \Psi_\beta}{\partial \beta} & -\frac{\partial \Psi_\beta}{\partial \gamma} \\ 0 & -\frac{\partial \Psi_\gamma}{\partial \gamma} \end{pmatrix} \quad \begin{pmatrix} 0 \\ 0 \end{pmatrix} \\ \begin{pmatrix} 0 & 0 \end{pmatrix} \quad \begin{pmatrix} -\frac{\partial \Psi_\alpha}{\partial \alpha} \end{pmatrix} \right]^{-1} \quad (3.144)$$

$\mathbf{B}$ is organized similarly. Since our goal is to obtain the sandwich estimate of variance for β, we need only look at the result of the matrix multiplications for the upper $p \times p$ entry of the sandwich estimate of variance. Since $\mathbf{A}$ is block diagonal, we need only look at $\mathbf{A}_{11}^{-1} \mathbf{B}_{11} \mathbf{A}_{11}^{-T}$.

Due to this simplification, we assume an independent correlation structure for a linear regression model since that allows us to certify results with commercial software that includes support for the sandwich estimate of variance for instrumental variables regression. Our approach will match the commercial software results, except in the situation where scalar adjustments may be made to the variance estimator.

Assume that we wish to model a continuous outcome using the identity link function

$$\mathbf{Y} = \beta_0 + \beta_1 \mathbf{x}_1 + \beta_2 \mathbf{x}_2 + \beta_3 \mathbf{x}_3 \quad (3.145)$$

with an exchangeable correlation PA-GEE model. However, we cannot observe $\mathbf{x}_3$. Instead, we observe $\mathbf{w}$ which is equal to $\mathbf{x}_3$ plus measurement error. In addition, we have an instrumental variable $\mathbf{s}$.

Since $\mathbf{x}_3$ is not observed, we can first fit a regression of $(1, \mathbf{x}_1, \mathbf{x}_2, \mathbf{s})$ on $\mathbf{w}$, and use the fitted values from the regression as a proxy for the unobserved variable. Using the data listed in section 5.2.6, the $\mathbf{A}_{11}$ matrix is estimated by

$$\mathbf{A}_{11} = \begin{pmatrix} -\frac{\partial \Psi_\beta}{\partial \beta} & -\frac{\partial \Psi_\beta}{\partial \gamma} \\ -\frac{\partial \Psi_\alpha}{\partial \beta} & -\frac{\partial \Psi_\alpha}{\partial \gamma} \end{pmatrix} \quad (3.146)$$

where the estimated submatrices for the derivative of the estimating equation

FURTHER DEVELOPMENTS AND APPLICATIONS

for the regression of interest are

$$-\frac{\partial \Psi_\beta}{\partial \beta} = \begin{pmatrix} 43.88 & -23.08 & -84.48 & -8.10 \\ -23.08 & 423.00 & 637.00 & 119.00 \\ -84.48 & 637.00 & 1578.00 & 216.00 \\ -8.10 & 119.00 & 216.00 & 40.00 \end{pmatrix} \quad (3.147)$$

$$-\frac{\partial \Psi_\beta}{\partial \gamma} = \begin{pmatrix} -92.48 & -338.55 & 164.70 & -32.45 \\ 1695.09 & 2552.66 & -131.00 & 476.87 \\ 2552.66 & 6323.53 & -330.96 & 865.58 \\ 476.87 & 865.58 & -39.77 & 160.29 \end{pmatrix} \quad (3.148)$$

and the estimated submatrices for the derivatives of the estimating equation of the instrumental variables regression are

$$-\frac{\partial \Psi_\gamma}{\partial \beta} = [0] \quad (3.149)$$

$$-\frac{\partial \Psi_\gamma}{\partial \gamma} = \begin{pmatrix} 423.00 & 637.00 & -32.69 & 119.00 \\ 637.00 & 1578.00 & -82.59 & 216.00 \\ -32.69 & -82.59 & 39.08 & -9.92 \\ 119.00 & 216.00 & -9.92 & 40.00 \end{pmatrix} \quad (3.150)$$

The $\mathbf{B}_{11}$ matrix is

$$\mathbf{B}_{11} = \begin{pmatrix} (\Psi_\beta)(\Psi_\beta) & (\Psi_\beta)(\Psi_\gamma) \\ (\Psi_\gamma)(\Psi_\beta) & (\Psi_\gamma)(\Psi_\gamma) \end{pmatrix} \quad (3.151)$$

where, for our data, the submatrices for the first row are estimated by

$$(\Psi_\beta)(\Psi_\beta) = \begin{pmatrix} 56.10 & 33.44 & -0.53 & 18.12 \\ 33.44 & 281.41 & 355.03 & 85.24 \\ -0.53 & 355.03 & 944.74 & 139.77 \\ 18.12 & 85.24 & 139.77 & 32.39 \end{pmatrix} \quad (3.152)$$

$$(\Psi_\beta)(\Psi_\gamma) = \begin{pmatrix} -14.59 & -35.63 & 2.38 & -5.75 \\ 5.55 & 55.51 & -12.91 & 0.92 \\ 55.51 & 143.07 & -33.85 & 13.42 \\ 0.92 & 13.42 & -5.20 & -0.74 \end{pmatrix} \quad (3.153)$$

and the submatrices for the second row are estimated by

$$(\Psi_\gamma)(\Psi_\beta) = \begin{pmatrix} -14.59 & 5.55 & 55.51 & 0.92 \\ -35.63 & 55.51 & 143.07 & 13.42 \\ 2.38 & -12.91 & -33.85 & -5.20 \\ -5.75 & 0.92 & 13.42 & -0.74 \end{pmatrix} \quad (3.154)$$

$$(\Psi_\gamma)(\Psi_\gamma) = \begin{pmatrix} 103.00 & 143.22 & -16.13 & 29.20 \\ 143.22 & 321.86 & -37.02 & 45.00 \\ -16.13 & -37.02 & 8.87 & -4.86 \\ 29.20 & 45.00 & -4.86 & 9.73 \end{pmatrix} \quad (3.155)$$

Using the estimated $\mathbf{A}_{11}$ and $\mathbf{B}_{11}$ matrices, the calculated sandwich estimate of variance for the instrumental variables regression is

$$\widehat{\mathbf{V}}_S(\beta) = \begin{pmatrix} .105844 & -.020986 & .003564 & .067679 \\ -.020986 & .077022 & .005081 & -.273977 \\ .003564 & .005081 & .01454 & -.110762 \\ .067679 & -.273977 & -.110762 & 1.68467 \end{pmatrix} \quad (3.156)$$

This variance estimator, even though it was calculated for a PA-GEE model with instrumental variables, is the same as a sandwich estimate of variance for instrumental variables regression. If we fit such a model in a commercial package, we obtain the results

```
IV (2SLS) regression with robust standard errors     Number of obs =      40
                                                     F(  3,    36) =  242.78
                                                     Prob > F      =  0.0000
                                                     R-squared     =  0.9506
                                                     Root MSE      =  2.3249

------------------------------------------------------------------------------
             |              Robust
           y |    Coef.    Std. Err.       t     P>|t|    [95% Conf. Interval]
-------------+----------------------------------------------------------------
           w |  4.00731    .3429365     11.69    0.000    3.311802    4.702817
          x1 |  1.878759   .2925399      6.42    0.000    1.28546     2.472057
          x2 |  3.141942   .1271036     24.72    0.000    2.884164    3.39972
       _cons |  .228422    1.368156      0.17    0.868   -2.546328    3.003172
------------------------------------------------------------------------------
Instrumented:  w
Instruments:   x1 x2 s
------------------------------------------------------------------------------
```

The commercial package (in this case, Stata) lists the sandwich estimate of variance as

$$\widehat{\mathbf{V}}_S^{\text{Stata}}(\beta) = \begin{pmatrix} .117605 & -.023317 & .003960 & .075198 \\ -.023317 & .085580 & .00565 & -.304419 \\ .003960 & .00565 & .016155 & -.123069 \\ .075198 & -.304419 & -.123069 & 1.87185 \end{pmatrix} \quad (3.157)$$

Mentioned as a possibility earlier, Stata does apply a documented scalar adjustment to the sandwich estimate of variance. The scalar adjustment is equal to $n/(n-p)$ where n is the number of observations and p is the number of covariates in the regression model. In this case, one can easily verify that

$$\widehat{\mathbf{V}}_S^{\text{Stata}}(\beta) = \frac{40}{40-4}\widehat{\mathbf{V}}_S(\beta) \quad (3.158)$$

While we did not list output for the PA-GEE model, the coefficients match the output from the commercial package.

To complete the illustration of the techniques, we should outline some of the formulas that are required for the calculation of the variance estimator. There is no commercial software that allows specification of these types of general models (though we saw an example of a specific member of the class of models).

The estimating equation Ψ_β for the regression model of concern is complicated by the covariate that is constructed from the fitted values of the

FURTHER DEVELOPMENTS AND APPLICATIONS

instrumental variable regression.

$$\left[\frac{\partial}{\partial \beta_j} \sum_{i=1}^{n}\left\{-\frac{1}{2}\left[y_i - (w_i\beta_1 + x_{2i}\beta_2 + x_{3i}\beta_3 + \beta_0)\right]^2\right\}\right] = [0]_{4\times 1} \quad (3.159)$$

with $j = 1, \ldots, 4$.

The most complicated part of calculating the sandwich estimate of variance is clearly the calculation of $-\partial\Psi_\beta/\partial\gamma_k^T$, since the construction of **w** involves the γ parameter vector. The remaining terms in the calculation of the sandwich estimate of variance are easily obtained using results from the separate regressions.

3.6.2 The PA-EGEE for GLMs

The PA-GEE discussed earlier is an application of the quasilikelihood to panel data. The partial derivatives of the quasilikelihood have score-like properties when the derivatives are in terms of β, but do not have these properties for the partial derivatives in terms of α. Nelder and Pregibon (1987) developed an extension to the quasilikelihood for which both partial derivatives have score-like properties. Hall and Severini (1998) subsequently utilized this approach to extend the GEE1 for GLMs. In the extension, it is assumed that the extended quasilikelihood may be written in the form

$$\mathcal{Q}^+(\mathbf{y}_i; \boldsymbol{\mu}_i, \boldsymbol{\alpha}) = \mathcal{Q}(\mathbf{y}_i; \boldsymbol{\mu}_i) + f_{i1}(\boldsymbol{\alpha}) + f_{i2}(\mathbf{y}_i) \quad (3.160)$$

ensuring that the partial derivative of the extended quasilikelihood with respect to β is the same as the partial derivative of the quasilikelihood with respect to β

$$\frac{\partial \mathcal{Q}^+}{\partial \beta} = \frac{\partial \mathcal{Q}}{\partial \beta} \quad (3.161)$$

Recall that the quasilikelihood associated with GLMs is given by

$$\mathcal{Q}(y; \mu) = \int^{\mu} \frac{y - \mu^*}{V(\mu^*)} d\mu^* \quad (3.162)$$

implying that

$$\frac{\partial \mathcal{Q}(y; \mu)}{\partial \mu} = \frac{y - \mu}{V(\mu)} \quad (3.163)$$

The deviance is then calculated as

$$D(y; \mu) = -2\{\mathcal{Q}(y; \mu) - \mathcal{Q}(y; y)\} \quad (3.164)$$

$$= -2 \int_y^{\mu} \frac{y - \mu^*}{V(\mu^*)} d\mu^* \quad (3.165)$$

The extended quasilikelihood may then be written in terms of these quantities as

$$\mathcal{Q}^+(y; \mu) = -\frac{1}{2}\ln\{2\pi\phi V(y)\} - \frac{1}{2}D(y; \mu)\frac{1}{\phi} \quad (3.166)$$

such that the estimate obtained by maximizing the extended quasilikelihood $\widehat{\beta}_{Q+}$ estimates the same population parameter as the estimate obtained maximizing the quasilikelihood $\widehat{\beta}_Q$. A proper likelihood is implied if there is a distribution in the exponential family, with the variance function specified for the extended quasilikelihood.

To illustrate the connection of the extended quasilikelihood to the models already examined, let us derive an estimating equation from the extended quasilikelihood for the exponential family using $V(\mu) = \mu$ and $a(\phi) = 1$ (the appropriate choices for the Poisson model). The extended quasilikelihood for this case is given by

$$Q^+(y;\mu) = -\frac{1}{2}\ln\{2\pi y\} + \int_y^\mu \frac{y-\mu^*}{V(\mu^*)a(\phi)}d\mu^* \qquad (3.167)$$

$$= -\frac{1}{2}\ln\{2\pi y\} + \int_y^\mu \frac{y-\mu^*}{\mu^*}d\mu^* \qquad (3.168)$$

$$= y\ln(\mu) - \mu - y(\ln(y) - 1) - \frac{1}{2}\ln\{2\pi y\} \qquad (3.169)$$

with an estimating equation $\Psi(\Theta) = \partial Q^+/\partial \mu = \mathbf{0}$ for $\Theta = \beta$ given by

$$\left[\left\{\frac{\partial Q^+}{\partial \beta_j} = \sum_{i=1}^n \left(\frac{y_i}{\mu_i}-1\right)\left(\frac{\partial \mu}{\partial \eta}\right)_i x_{ji}\right\}_{j=1,\ldots,p}\right]_{p\times 1} = [\mathbf{0}]_{p\times 1} \qquad (3.170)$$

Equation 3.170 matches the specific derivation of the estimating equation for the likelihood-based Poisson model given in equation 2.29. The estimating equations for the two approaches match even though the extended quasilikelihood in equation 3.169, implied by assuming a variance function from the Poisson distribution, differs (in the normalizing term) from the Poisson log-likelihood given (in terms of $\mathbf{x}\beta$) in equation 2.28.

Solving the partial derivative with respect to $\boldsymbol{\alpha}$ would be a straightforward approach to deriving the estimating equations from the extended quasilikelihood. However, there are two problems with this approach. First, solving the partial derivative is difficult, and second, the resulting estimator is biased. Utilizing the same decomposition of the variance as was used for the PA-GEE model, we require a matrix

$$\int_0^1 s\mathrm{D}(\mathrm{V}(\mu_{it}))^{-1/2}\{t(s)\}\frac{\partial \phi^{-1}\mathbf{R}(\boldsymbol{\alpha})}{\partial \alpha_j}\mathrm{D}(\mathrm{V}(\mu_{it}))^{-1/2}\{t(s)\}ds \qquad (3.171)$$

where the elements of the matrix are functions of $\boldsymbol{\alpha}$ that depend on another integral. For our purposes, it is enough to understand that this approach is computationally vexing due to numeric integration of functions with end point singularities. The solution (assuming we can get to one) leads to biased estimates. However, we emphasize that we could, in fact, proceed with solving these integrals out of a desire to fit a *true* extended quasilikelihood model.

Alternatively, the integral (equation 3.171) may be approximated using a first-order Taylor series expansion, providing the PA-EGEE given by

$$\Psi(\beta, \alpha) = (\Psi_\beta(\beta, \alpha), \Psi_\alpha(\beta, \alpha)) \qquad (3.172)$$

$$= \begin{pmatrix} \sum_{i=1}^{n} \mathbf{x}_{ji}^T D\left(\dfrac{\partial \mu_i}{\partial \eta}\right) (V(\mu_i))^{-1} \left(\dfrac{\mathbf{y}_i - \mu_i}{a(\phi)}\right) \\ \sum_{i=1}^{n} \left[-(\mathbf{y}_i - \mu_i)^T \dfrac{\partial V(\mu_i)^{-1}}{\partial \alpha}(\mathbf{y}_i - \mu_i) \right. \\ \left. + \mathrm{tr}\left(V(\mu_i)\dfrac{\partial V(\mu_i)^{-1}}{\partial \alpha}\right) \right] \end{pmatrix} \qquad (3.173)$$

$$V(\mu_i) = D(V(\mu_{it}))^{1/2}\, \mathbf{R}(\alpha)\, D(V(\mu_{it}))^{1/2} \qquad (3.174)$$

Hall (2001) points out that we can make use of the fact that

$$V(\mu_i)\dfrac{\partial V(\mu_i)^{-1}}{\partial \alpha} = -\dfrac{\partial V(\mu_i)}{\partial \alpha} V(\mu_i)^{-1} \qquad (3.175)$$

in order to avoid the need to differentiate $V(\mu_i)^{-1}$ in calculating the estimating equation.

To use this model in practice requires programming since there is currently no support for this class of models in existing software packages. Choosing between fitting a PA-EGEE model and a GEE2 model is usually based on the focus of the analysis and the reasonableness of treating the two estimating equations as orthogonal. In general, PA-EGEE compared to a similar GEE2 model provides smaller standard errors for β (because of the orthogonality assumption) and a less accurate estimate of the dispersion ϕ.

3.6.3 The PA-REGEE for GLMs

Following ideas introduced for *robust regression* to allow for models resistant to outliers in the data, Preisser and Qaqish (1999) generalized the concepts for PA-GEE models. The resistant PA-GEE for $\beta_{p\times 1}$ is given by*

$$\Psi(\beta) = \sum_{i=1}^{n}\sum_{t=1}^{n_i} \left(\dfrac{\partial \mu}{\partial \eta}\right)_{it} [V(\mu_i)]^{-1} \left(\mathbf{w_i}\dfrac{\mathbf{y}_i - \mu_i}{a(\phi)} - \mathbf{c}_i\right) = [\mathbf{0}]_{p\times 1} \qquad (3.176)$$

where the usual PA-GEE is a special case wherein $\mathbf{w_i}$ is an $n_i \times n_i$ identity matrix (for all i) and $\mathbf{c}_i$ is an $n_i \times 1$ vector of zeros (for all i). The estimating equation for the association parameters α, due to Liang and Zeger (1986), is given in equation 3.15 (using moment estimates based on Pearson residuals). In order to use the estimating equation using ALR in equation 3.79, or the estimating equation from the PA-EGEE model, one would have to first work out the required robust formulae changes to those estimating equations. The changes to the moment estimates for the estimating equation for PA-GEE are given later in this section.

---

* We denote the resistant PA-GEE as PA-REGEE, whereas the cited authors use REGEE.

This presentation, like that for PA-GEE in equation 3.14, assumes that the variance of the outcomes may be written

$$V(\boldsymbol{\mu}_i) = \left[D(V(\mu_{it}))^{1/2} \, \mathbf{R}(\boldsymbol{\alpha})_{(n_i \times n_i)} \, D(V(\mu_{it}))^{1/2} \right]_{n_i \times n_i} \quad (3.177)$$

In general, $\mathbf{w}_i$ is a diagonal matrix of observation weights and $\mathbf{c}_i$ is a vector of constants ensuring that the estimating equation is unbiased.

The Mallows class of weights determines observation weights as a function of the values of the covariates only. The Schweppe class of weights determines weights as a function of the outcomes. The basic idea of resistant estimation is to investigate the influence of the observations and then downweight influential data so that a more even contribution to the estimation is obtained for each observation or panel. As seen in Chapter 4, influence may be measured per observation or per panel. Therefore, we may apply the downweighting based on either approach. The theoretical justification of the approach is discussed in the cited article as well as Carroll and Pederson (1993) for the case of models with binomial variance and the logit link function.

For the Mallows class of weights, we have $\mathbf{c}_i = \mathbf{0}$ for all i, and we need not make any further assumptions past those for PA-GEE. Following the fit of a PA-GEE model, the Mallows weights may be determined through an investigation of the influence and then a new PA-REGEE model fit with the weights determined in the previous step. Even if you have access to a statistical package that allows weights, you may not be able to use it. First, some statistical packages require that weights be constant within panel (limiting you to panel-level downweighting), and second, the statistical package may not apply the weights in the desired manner for the calculation of the moment estimates. Check the documentation of your preferred software package to see if this can be done. Otherwise, programming is required.

Preisser and Qaqish made prenatal care data available* that we shall analyze. The data include 137 observations (patients) for 42 doctors. The outcome is whether the patient is bothered by urinary incontinence. Clustered by doctor `doct_id`, the covariates included in the model are `female`, the gender of the patient; `age`, the age in decades; `dayacc`, a constructed daily number of leaking accidents based on the reported number of accidents per week; `severe`, whether the accidents are severe; and `toilet`, the average number of times the patient uses the toilet per day.

The PA-GEE fit for the data is given by

```
GEE population-averaged model           Number of obs       =        137
Group variable:              doct_id    Number of groups    =         42
Link:                          logit    Obs per group: min  =          1
Family:                     binomial                   avg  =        3.3
Correlation:            exchangeable                   max  =          8
                                        Wald chi2(5)        =      30.16
Scale parameter:                   1    Prob > chi2         =     0.0000
```

* http://www.phs.wfubmc.edu/data/uipreiss.html

```
------------------------------------------------------------------------
   bothered |      Coef.   Std. Err.       z     P>|z|   [95% Conf. Interval]
------------+-----------------------------------------------------------
     female | -.7730688    .6012588    -1.29     0.199   -1.951514    .4053767
        age | -.6556766    .575984     -1.14     0.255   -1.784585    .4732313
     dayacc |  .3972632    .0926325     4.29     0.000    .2157068    .5788196
     severe |  .8027313    .3530613     2.27     0.023    .1107439   1.494719
     toilet |  .1059107    .0841537     1.26     0.208   -.0590274    .2708489
      _cons | -3.035959   1.111234     -2.73     0.006   -5.213939   -.8579799
------------------------------------------------------------------------
```

The estimated exchangeable correlation is 0.1013.

Note that the PA-REGEE model generalizes the estimating equation to downweight for influential observations. The moment estimators must be downweighted as well. The dispersion parameter is estimated using

$$\widehat{\phi} = \frac{1}{n^* - p} \sum_{i=1}^{n} \sum_{t=1}^{n_i} \widehat{r}_{it}^{*\,2} \qquad (3.178)$$

where

$$\widehat{r}_{it}^{*} = \frac{y_{it} - \widehat{\mu}_{it} - c_{it}}{V(\mu_{it})} \qquad (3.179)$$

$$n^* = \sum_{i=1}^{n} \sum_{t=1}^{n_i} w_{it}^2 \qquad (3.180)$$

Specifying panel-level Mallows-class downweights results in

```
GEE population-averaged model            Number of obs      =       137
Group variable:                doct_id   Number of groups   =        42
Link:                            logit   Obs per group: min =         1
Family:                       binomial                  avg =       3.3
Correlation:              exchangeable                  max =         8
                                         Wald chi2(5)       =     32.15
Scale parameter:                     1   Prob > chi2        =    0.0000

------------------------------------------------------------------------
   bothered |      Coef.   Std. Err.       z     P>|z|   [95% Conf. Interval]
------------+-----------------------------------------------------------
     female | -.8275123    .6234338    -1.33     0.184   -2.04942     .3943956
        age | -.2153152    .5897718    -0.37     0.715   -1.371247    .9406162
     dayacc |  .3800309    .0889533     4.27     0.000    .2056856    .5543762
     severe |  .9275332    .3542698     2.62     0.009    .2331773   1.621889
     toilet |  .0677876    .0821533     0.83     0.409   -.0932299    .2288051
      _cons | -3.141697   1.124495     -2.79     0.005   -5.345666   -.9377271
------------------------------------------------------------------------
```

Utilizing the Schweppe class of weights is more complicated since we must determine the vector of constants $\mathbf{c}_i$—hence ensuring unbiasedness for the estimating equation.

3.7 Missing data

Techniques for dealing with missing data are steadily gaining recognition and there is currently active research aimed at developing new techniques for specific modeling situations. This subject is far larger in scope than we can detail within the limits of our text. Our introduction here is designed to introduce the reader to the topic and to outline some of the techniques that have been successfully applied—especially for the case of dropouts in longitudinal data studies. In the subsequent chapter we present techniques for assessing missing data together with formal tests of the MCAR assumption.

We anticipate that commercial software packages will add sophisticated techniques for modeling panel data with missing observations. However, these additions will not be turnkey solutions since the analyst will be required to make major modeling decisions as to the nature and assumptions underlying the applied techniques. This section outlines those assumptions and explains the motivations and implications of various types of missing data.

Throughout the text, we have thus far implicitly assumed that the data we analyze are complete. However, this is often not true in practice. The figure below illustrates various patterns of missing data.

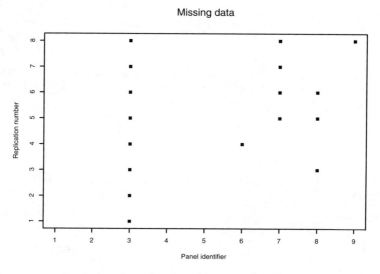

Squares mark missing data for the response variable in a dataset with 9 panels and 8 repeated measures per panel.

In the figure above, the missing data patterns are identified as:
- **Complete:** Panels 1, 2, 4, and 5 are complete panels where there are no missing data. These panels provide complete information for the model.
- **Panel nonresponse:** Panel 3 has no observations; all replications are missing. This panel provides no information for the model.

- **Item nonresponse:** Panels 6 and 8 have some missing data. These panels provide incomplete information for the panels.
- **Dropout:** Panels 7 and 9 have a special type of item nonresponse where once an observation is missing for the panel, the rest of the observations are also missing. These panels provide incomplete information for the panels.

In viewing the patterns of the missing data, we are concerned with whether that pattern is random or monotone. To investigate the process that generates the missing observations, we partition the outcomes into

$$\mathbf{Y} = \text{complete data} \qquad (3.181)$$
$$\mathbf{Y}_o = \text{observed data} \qquad (3.182)$$
$$\mathbf{Y}_m = \text{missing data} \qquad (3.183)$$

and we construct an indicator matrix $\mathbf{M}$ for the missingness of observations where the elements of the matrix are defined as

$$M_{it} = \begin{cases} 1 & Y_{it} \text{ is missing} \\ 0 & Y_{it} \text{ is observed} \end{cases} \qquad (3.184)$$

Our goal is to investigate the joint distribution $f_{\mathbf{Y},\mathbf{M}}$ insofar as we are interested in knowing whether the distribution of the missing data $f_{\mathbf{M}}$ is independent of the outcomes. Essentially we want to know if $f_{\mathbf{M}|\mathbf{Y}} = f_{\mathbf{M}}$. We define several useful terms based on probabilities for characterizing the missing data.

If $P(\mathbf{M}|\mathbf{Y}) = P(\mathbf{M})$ for all $\mathbf{Y}$, then $\mathbf{M}$ is independent of the observed outcomes $\mathbf{Y}_o$ and the missing outcomes $\mathbf{Y}_m$. In this case, the process for missing data is called missing completely at random, or MCAR. Rotnitzky and Wypij (1994) explain that the MCAR assumption means the process that generates missing data is independent of the observed and unobserved data values. In such a case, the standard techniques we have discussed provide valid inferences.

If $P(\mathbf{M}|\mathbf{Y}) = P(\mathbf{M}|\mathbf{Y}_o)$ for all $\mathbf{Y}_m$, then $\mathbf{M}$ is independent of $\mathbf{Y}_m$. For this case, the process for missing data is called missing at random, or MAR. Rubin (1976) points out that valid inference is obtained from likelihood-based models that ignore the missing data mechanism when the nonresponse depends on the observed data; but the nonresponse mechanism is still independent of the unobserved data.

If $P(\mathbf{M}|\mathbf{Y})$ depends on the missing outcomes $\mathbf{Y}_m$, the missing data are called informatively missing data or nonignorable nonresponse.

In a catalog of analysis techniques, we can partition our data into complete and incomplete cases. Imputation is typically the first approach used to handle missing data. In this technique, missing values are replaced with some imputed value from the data. This is a simple technique, but requires assumptions on how to impute the values. The validity of the results of imputation are directly tied to the assumptions used in imputing the missing data.

We discuss the example data of the classification of asthma among white children from Steubenville, Ohio. This example is also used as the motivating example in Rotnitzky and Wypij (1994). The data consists of 1419 children (706 boys and 713 girls) where the classification of asthma status is recorded for each child at age 9 and age 13. There are 149 missing classifications for boys at age 13, and 123 missing classifications for girls at age 13. The data are summarized as

		BOYS			
		Asthma at age 13			
		No	Yes	Missing	Total
Asthma	No	514	15	145	674
at age 9	Yes	6	22	4	32
	Total	520	37	149	706
		GIRLS			
		Asthma at age 13			
		No	Yes	Missing	Total
Asthma	No	561	13	115	689
at age 9	Yes	3	13	8	24
	Total	564	26	123	713

These data have $i = 1, \ldots, 1419$ and $t = 1, 2$ where the repeated observations are for age 9 and age 13. There are 1147 complete panels and 272 dropout panels for which the outcome is unobserved at age 13; thus, there are $1147(2) + 272(1) = 2566$ observations. We assume that the outcomes follow a logistic model where the covariates include a constant, an indicator variable for gender, and an indicator variable for age 13.

If we fit a logistic regression model (a PA-GEE model assuming independence of the repeated observations) to the data ignoring any mechanism for the missing data, we obtain the following results

```
GEE population-averaged model              Number of obs      =      2566
Group variable:                       id   Number of groups   =      1419
Link:                              logit   Obs per group: min =         1
Family:                         binomial                  avg =       1.8
Correlation:                 independent                  max =         2
                                           Wald chi2(2)       =      7.27
Scale parameter:                       1   Prob > chi2        =    0.0264

Pearson chi2(2566):              2568.36   Deviance           =    955.94
Dispersion (Pearson):           1.000921   Dispersion         =  .3725423
```

```
------------------------------------------------------------------------
       y  |     Coef.   Std. Err.      z    P>|z|    [95% Conf. Interval]
----------+-------------------------------------------------------------
   gender |  .3750074   .1902226     1.97   0.049    .0021778   .7478369
    age13 |   .351794   .1882782     1.87   0.062   -.0172244   .7208124
    _cons | -3.394797   .1758445   -19.31   0.000   -3.739446  -3.050148
------------------------------------------------------------------------
```

The validity of the inferences we draw for fitted models on incomplete data is a function of whether the mechanism generating the missing data is ignorable. Nonignorable missing data result in biased coefficient estimates.

We can hypothesize many reasons for the missing data in the example presented. We can assume that the missing data are related to the asthma status such that those without asthma at age 13 are always observed; but those with asthma have some probability of not being observed. Under this assumption, the imputed complete table of responses would be

		BOYS		**GIRLS**	
		Asthma at age 13		Asthma at age 13	
		No	Yes	No	Yes
Asthma	No	514	160	561	118
at age 9	Yes	6	26	3	21

Missing data all assigned as asthmatics.

Under this imputation, the coefficient table for the independence model is

```
------------------------------------------------------------------------
       y  |     Coef.   Std. Err.      z    P>|z|    [95% Conf. Interval]
----------+-------------------------------------------------------------
   gender |  .3588865   .1157747     3.10   0.002    .1319722   .5858008
    age13 |   1.98622   .1505188    13.20   0.000    1.691209   2.281232
    _cons | -3.379309   .1522291   -22.20   0.000   -3.677673  -3.080946
------------------------------------------------------------------------
```

Results from this assumption show how our estimate of the coefficient on age13 is downward biased when we analyze only the observed data.

Instead of assuming that the asthmatics might not respond, we can assume that it is the non-asthmatics who might not respond. In this case, the imputed complete data table is

		BOYS		**GIRLS**	
		Asthma at age 13		Asthma at age 13	
		No	Yes	No	Yes
Asthma	No	659	15	676	13
at age 9	Yes	10	22	11	13

Missing data all assigned as non-asthmatics.

The coefficient table for the independence model under this imputation is then

```
         y |      Coef.   Std. Err.       z     P>|z|     [95% Conf. Interval]
-----------+----------------------------------------------------------------
    gender |   .3464105   .1896561     1.83     0.068    -.0253087    .7181297
     age13 |   .1230778   .1877312     0.66     0.512    -.2448685    .4910241
     _cons |  -3.378212   .1748608   -19.32     0.000    -3.720933   -3.035491
```

Under this assumption, our estimated coefficient on the `age13` variable, using only the observed data, is biased upward instead of downward. Clearly, we can make many other assumptions about the nature of the mechanism driving the missingness of data. Under some assumptions, an analysis using only the observed data will not differ significantly from the (unknown) complete data.

A second approach to analyzing data with missing values is another form of complete case analysis. In this approach, we drop the incomplete cases and generate weights for the complete cases to address bias induced by the missing data process. This can be difficult to do with existing software unless the software supports user-defined weights.

If we assume that the data are MAR, we can calculate probabilities of non-response to construct a probability weighted estimating equation. Assuming that the data are missing as a function of gender and the observed outcome at age 9, the probability of nonresponse for boys who were classified as asthmatics at age 9 is $145/674 = .215$, for boys who were not classified as asthmatics at age 9 is $4/32 = .125$, for girls who were classified as asthmatics at age 9 is $115/689 = .167$, and for girls who were not classified as asthmatics at age 9 is $8/24 = .333$. Fitting this weighted model results in

```
GEE population-averaged model               Number of obs      =      2566
Group variable:                    id       Number of groups   =      1419
Link:                           logit       Obs per group: min =         1
Family:                      binomial                      avg =       1.8
Correlation:              exchangeable                     max =         2
                                            Wald chi2(2)       =     13.24
Scale parameter:                    1       Prob > chi2        =    0.0013
```

```
                      (standard errors adjusted for clustering on id)
-------------------------------------------------------------------------
             |                Semi-robust
           y |      Coef.     Std. Err.      z    P>|z|    [95% Conf. Interval]
-------------+-----------------------------------------------------------
      gender |   .4119338     .2335005     1.76   0.078    -.0457187    .8695863
       age13 |    .365495     .1161503     3.15   0.002     .1378446    .5931454
       _cons |  -3.418043     .1872194   -18.26   0.000    -3.784986    -3.0511
-------------------------------------------------------------------------
```

A third approach to analyzing data with missing values in a PA-GEE model is to assume that the process generating the missing data admits this estimation and proceeds with an incomplete analysis. In this approach, all complete observations (regardless of whether the panel is complete) are included in the analysis. The PA-GEE model actually requires a special case of the MCAR assumption; we assume that $P(M|Y,X) = P(M|X)$ for all Y. Conditional on the covariates, M is independent of the observed outcomes Y_o as well as the missing outcomes Y_m. Further, PA-GEE modeling is appropriate if a dataset has missing values generated from a dropout process, if the data are MAR, and if the parameters of the dropout process are distinct from the parameters of interest. This assumption is analyzed by Shih (1992) where he outlines the necessary conditions subject to distinct parameters.

The most commonly studied pattern of missing data relates to dropouts. In fact, this is a common outcome in many health related studies. Imagine a health study in which patients are randomized to a treatment drug or to a placebo. It is reasonably expected (and common) that those patients assigned to the placebo may stop participating after several observations when there is no change in their status. Likewise it is sensical (and common) that those patients assigned to the treatment drug may be susceptible to a side effect that causes their participation to stop at some point in the study. In fact, these types of dropouts are sometimes designed into a health study in order to safeguard the participants.

In modeling dropouts, the basic idea is to include a model for the complete cases and a model for the dropouts. Various interactions are hypothesized for considering the joint distribution of these two models. Typically, such investigations result in likelihood-based techniques not covered in this text. Interested readers can see Little (1995) for an excellent example.

Robins, Rotnitzky, and Zhao (1995) present another approach for modeling dropouts; see also Rotnitzky and Robins (1995). The authors present a weighted estimating equation resulting in valid unbiased estimates under the assumption that the probability that an observation is missing depends only on the past values of the covariates and outcomes.

The approach amounts to a weighting scheme based on the inverse probability of censoring that extends the GEE class of models to MAR-classified data.

It is important to note that we have switched the notation from the original paper. The authors' discussion centers on an observation being uncensored (not missing) $R_{it} = 1$; our present discussion centers on an observation being missing $M_{it} = 0$.

The authors suppose that the response probabilities are given by

$$\lambda_{it} = P(M_{it} = 0 | M_{it-1} = 0, X_{i1}, \ldots, X_{it}, Y_{i1}, \ldots, Y_{it}) \quad (3.185)$$

This equation says that the conditional probability that the itth observation is not missing given that the previous observation is not missing and given all of the covariates and outcomes up to time t is equal to λ_{it}. It is assumed that these conditional probabilities are known up to q unknown parameters. The basic idea is then to model this conditional probability by a logistic regression. Fitted values are then used as the weights in the GEE or other model.

The PA-GEE is generalized for this inverse probability weighting as:

$$\Psi(\boldsymbol{\beta}, \boldsymbol{\alpha}) = (\Psi_{\boldsymbol{\beta}}(\boldsymbol{\beta}, \boldsymbol{\alpha}), \Psi_{\boldsymbol{\alpha}}(\boldsymbol{\beta}, \boldsymbol{\alpha})) \quad (3.186)$$

$$= \begin{pmatrix} \sum_{i=1}^{n} \mathbf{x}_{ji}^{T} D\left(\frac{\partial \boldsymbol{\mu}_i}{\partial \boldsymbol{\eta}}\right) (V(\boldsymbol{\mu}_i))^{-1} \left(\frac{\mathbf{y}_i - \boldsymbol{\mu}_i}{a(\phi)}\right) \\ \sum_{i=1}^{n} \left(\frac{\partial \boldsymbol{\xi}_i}{\partial \boldsymbol{\alpha}}\right)^{T} \mathbf{H}_i^{-1} (\mathbf{W}_i - \boldsymbol{\xi}_i) \end{pmatrix} \quad (3.187)$$

$$V(\boldsymbol{\mu}_i) = D(\lambda_{it}^{-1}(1 - M_{it})) D(V(\mu_{it}))^{1/2} \mathbf{R}(\boldsymbol{\alpha}) D(V(\mu_{it}))^{1/2} \quad (3.188)$$

where the diagonal matrix of weights $D(\lambda_{it}(1 - M_{it}))$ are formed from the fitted values of the logistic regression. Readers interested in applying these techniques will have to program the necessary components since most software packages do not support individual level weights. The documentation for SUDAAN indicates that it supports specified observation-level weights. As we previously alluded, this technique is not limited to PA-GEE models.

There are, of course, additional likelihood-based modeling approaches to missing data. Fitzmaurice, Laird, and Lipsitz (1994) present a study with balanced models where missing data are classified as MAR. In this approach, the focus is on marginal models where associations are based on conditional log-odds ratios. The approach relies on the EM algorithm, see Dempster, Laird, and Rubin (1977), and requires substantial programming on the part of the interested analyst due to the lack of commercial software support. Other approaches for specific types of missing data are addressed in Diggle and Kenward (1994), Heyting, Tolboom, and Essers (1992), and Little and Rubin (1987).

3.8 Choosing an appropriate model

The previous chapter outlined the derivation of likelihood-based models and illustrated model construction and assumptions. Likewise, this chapter illustrates the techniques and construction of GEE models. Given a panel dataset, which model should an analyst choose to estimate? The answer is driven by a combination of factors: the scientific questions of interest, the size and nature of the panel dataset, and the nature of the covariates.

If the scientific questions of interest center on the individual effects of covariates on the response variable, then a subject-specific likelihood-based model or a subject-specific GEE model is most appropriate. Population-averaged models are not appropriate in this case and there is no way to alter the interpretation of the fitted coefficients to allow interpretation in a subject-specific manner. Valid likelihood-based models to address subject-specific hypotheses include unconditional fixed-effects models, conditional fixed-effects models, and random-effects models.

On the other hand, if the scientific questions center on the marginal effects of covariates, then a population-averaged model is appropriate; subject-specific models are not appropriate. The beta-binomial model is an example of this type of valid likelihood-based marginal model. Appropriate GEE models include the PA-GEE (using either moment estimators or ALR), PA-EGEE models, PA-REGEE models, or GEE2 models.

In a longitudinal dataset, we imagine data where the outcome is whether an individual student attends an optional study session. In these data, there are several study sessions over the semester in which we collect data. One covariate is the student's age. A second covariate is an indicator of whether the student failed the quiz immediately preceding the study session.

If we want to answer the question of whether the attendance depends on the age of the student, then a population-averaged model is appropriate. If we want to answer the question of whether the probability of attending the study session changes when an individual learns he or she is failing the course, then the subject-specific model is appropriate. In this example, we would fit both types of models in order to answer the scientific questions of interest.

Now imagine that the data are collected for a single optional study session and the panels are identified by the course in which the student is enrolled. Instead of a longitudinal dataset, we have a panel dataset; there are no repeated measurements on the individual students. We fit a population averaged model to answer the question of whether the probability of attending the study session depends on the age of the student. This model does not take advantage of repeated measurements even if such information exists in the data. Thus there is no change in the manner in which we interpret the coefficients.

To answer the question of whether the probability of attending the session depends on whether an individual student has failed the previous quiz, we can fit a subject-specific model. However, in this case the interpretation of the coefficient is more difficult. The coefficient's interpretation is based on a *change* in whether the student has failed the preceding quiz, and we have no such observations.

In general, population-averaged models are most appropriate for assessing changes in covariates that are constant within the panel identifier. In contrast, the population-averaged interpretation addresses the question of whether the probability of attending the study session depends on whether the previous quiz was failed, averaging over all students.

Both subject-specific and population-averaged GEE models depend on the availability of a sufficient number of panels in the dataset to be analyzed. A fixed-effects model is the most appropriate model if there are a small number of panels.

We cannot include covariates with values that are constant within panels in a fixed-effects (unconditional or conditional) model. Such covariates are called *panel level covariates*. Even if our focus is on interpreting the subject-specific effects in our experimental data, we cannot separate the effects of panel level covariates from the fixed effect—they are collinear.

Assuming that a population averaged model is appropriate, there is still a choice between using the moment estimators of the correlation matrix or the ALR approach estimating correlations based on log odds ratios. We recommend using ALR when the data are binary, especially if the focus of the analysis includes interpretation of the correlation coefficients. If the data are not binary and the focus of the analysis includes interpretation of the correlations, then a GEE2 model is preferred over a GEE1 model. For example, the PA-EGEE model, compared to a similar GEE2 model, provides smaller standard errors for β and a less accurate estimate of the dispersion ϕ. The smaller standard errors are a result of assuming orthogonality of the estimating equation for the regression coefficients and the estimating equation for the correlation parameters.

Within a class of GLMs for correlated data, the initial choice of the variance function is driven by the range and nature of the outcome variable. The binomial variance $\mu(1-\mu)$ is preferred if the outcome is binary. The Poisson variance μ is preferred if the outcomes represent counts of events. The Gaussian variance 1, gamma variance μ^2, or inverse Gaussian variance μ^3 may be used if the outcome is (effectively) continuous. Of course, the gamma and inverse Gaussian are most appropriate to use when the response consists of positive valued continuous numbers. However, once the initial variance is chosen, residual analysis is used to investigate the fit of the data for the chosen function. In Chapter 4 we illustrate an analysis of the variance function, which includes these steps of the analysis (see section 4.2.3).

Likewise, the initial choice of the link function for a particular model is usually chosen based on the range of the outcome variable. In most cases the canonical link is used. Whether we choose the canonical link or some other link usually has no effect on the outcome of the analysis, but can affect the calculation of the sandwich estimate of variance. This comes down to whether software uses the expected or observed information matrix for the construction of the sandwich estimate of variance. In the case of the canonical link, the two calculations are equivalent. Most software implements the expected Hessian specified by Liang and Zeger (1986). Shah et al. (1997) document the options available in the SUDAAN package whereby users can specify either approach. Hardin and Hilbe (2001) illustrate the relationship and derivation of both variance estimate constructions.

The above distinction is not as clear in the case of choosing between various random-effects models. For any given model, one can hypothesize any desired distribution for the random effects. Hopefully, the choice of the distribution is based on some scientific knowledge of the process, but this need not be the case. As long as the distribution supports a variety of shapes, depending on the distribution parameters, the model may be appropriate. In most cases the choice of the distribution for the random effects is driven by the integrability of the resulting likelihood for a panel. Residual analysis can help to distinguish a good model from a poor model. Standard model criterion such as Akaike's information criterion or the deviance statistic can be used to choose between a small collection of possible models.

3.9 Summary

In this chapter we have illustrated various approaches to building models from GEE in order to fit panel data. In so doing, we examined both the GEE1 and GEE2 methods. Within the GEE1 framework, the most well-known approach is that of Liang and Zeger (1986). There are many software packages that offer software support of these models. In general, though, the estimation of the association parameters is secondary to the analysis of interest, and no standard errors are reported. For the specific case of binomial models, the ALR technique is a subset of GEE1. This approach generally produces better estimates of the association parameters. Supporting software typically includes estimates of standard errors. There is excellent support for this technique from commercial software.

The PA-EGEE approach was the third GEE1 technique examined. This technique, like ALR, specifies a more formal estimating equation for the association parameters; however, there is at present no commercial software support.

In a situation where we only have access to software without support for one of the alternative GEE1 approaches, we can still fit the alternate models if the software allows us to both specify fixed correlation matrices and to limit the number of iterations to one. In this way, we can solve the association parameter estimating equation ourselves, build the working correlation matrix, and then use the PA-GEE software with our specified matrix to iterate once for an update to the current parameter estimates.

In addition, we reviewed a resistant GEE1 method for building GEE1 models that are resistant to outliers. There were two different approaches to building and specifying the downweights for the model discussed. There is currently no software support for this approach. We currently have to engage in programming in order to fit these adapted models.

The GEE2 approach differs from that of the GEE1 in that the estimating equation for the association parameters is not assumed to be orthogonal to the estimating equation for the regression coefficients. The regression coefficients β for the PA-EGEE model estimate the same population parameter as is estimated by the PA-GEE model, even though the two approaches are not numerically the same since the moment estimates of α differ. The estimating equation for the association parameters for the PA-EGEE model is the same as that for the GEE2 model when the GEE2 model assumes a Gaussian distribution for the random component—see Hall (2001). Note that GEE2 models are less likely to converge than are similar GEE1 models due to the increased complexity of the model.

It should be noted that many papers and researchers make reference to GEE. In nearly all cases, this reference is to the PA-GEE model described by Liang and Zeger (1986) where the association parameters are estimated via the Pearson residuals. There are several reasons for the popularity of these models. The original description of the PA-GEE models included illustrations of how to alter the IRLS algorithm so that these models could be estimated. The ease with which one could do this led to a large number of adoptions by various software packages and, more often, by individual users of software packages. The reader should note that this text attempts to clarify the widening field of GEE models by defining a taxonomy. This taxonomy has not been used prior to this text, and may not be adopted in future articles dealing with the subject. In many cases, you will read journal articles that make clear the model of interest in context. We have tried to adopt individual notational conventions for identifying GEE models into our presentation wherever possible and to point out differences where our notation differs. Regardless of an acceptance of our taxonomy by researchers in general, we believe that our notation does allow a clear distinction within this text to differentiate the models under discussion.

Ziegler, Kastner, Grömping, and Blettner (1996) recommend that analysts limit analyses to GEE1 models only when the panel sizes are less than or equal to 4 (and there are at least 30 panels). This recommendation follows from the simulation results of Liang and Zeger (1986) where they showed only small gains in efficiency for the PA-GEE models. The advice includes mention of at least 30 panels. This is now a standard rule of thumb for applying asymptotically justified estimators. With small panel sizes, one should also compare results from the PA-GEE models with the independence model.

In choosing an appropriate model, we acknowledge that software makes it possible for analysts to fit any number of models with relative ease. While there is some misuse of software for GLMs by analysts fitting every link and variance function, this is a poor use of software. There is an even greater opportunity for this type of model-hunting expedition with panel data since there are so many more possible models that might be estimated. Data analysis and model inference starts with an analysis of the scientific questions of interest that ultimately leads to a small collection of models to be estimated. Estimating all possible models (because software allows it) is scientifically irresponsible, and rarely leads to sound analysis.

There are two main approaches to handling missing observations in panel data analysis. The first is imputation. This technique fills in missing data with values imputed from the observed data. Various techniques form these proxy observations based on parametric or nonparametric assumptions. The second technique is to embed in the GEE another model for the mechanism that generates the missing observations. Either technique includes assumptions about the nature of the missing observations that must be carefully considered by the analyst. Our discussion illustrated the effects on estimators of various as-

SUMMARY

sumptions, presented the characteristics of different types of missing data, and motivated the need for sophisticated techniques to resolve the bias associated with models relying on the MCAR assumption.

We have made a concerted effort to compare and contrast models by focusing on their estimation algorithms and calculation. We believe that understanding the calculation of the models offers insight into the properties of estimators for different types of data and can illuminate the situations that lead to numeric difficulties. Finally, our focus on the algorithms and various choices for ancillary parameter estimators clears up the frustration many of us feel when comparing output across different software packages.

3.10 Exercises

1. This chapter focused a lot of attention on two estimators for the dispersion parameter in the PA-GEE model. Choose one of these estimators and present a case why you think it is the better choice.

2. Using your preferred software, determine which estimator of ϕ your software uses by fitting the small dataset illustrating the two competing estimators. Determine if your preferred software has options for using either approach.

3. Pan (2001b) introduced an alternate calculation of the modified sandwich estimate of variance. Explain the difference for Pan's formulation from the usual sandwich estimate of variance. For large samples, do you think there will be a significant difference from the usual calculation?

4. Explain why the sandwich estimate of variance for the PA-GEE model results in standard errors for the regression coefficients that are robust to misspecification of the hypothesized correlation structure.

5. Using the ship accident data, fit a PA-GEE model assuming a stationary(1) correlation structure. Interpret the regression coefficients as incidence rate ratios (IRRs). If your software prints out coefficients, then calculate the IRRs and standard errors using the delta method.

6. Discuss the motivations for the preference of independence models over more complicated models that include parameters for correlated data.

7. For the class of GLMs show that the observed Hessian is equal to the expected Hessian when the GLM is constructed with the canonical link. Discuss what this implies for the sandwich estimate of variance for PA-GEE models.

8. Data are collected for a longitudinal study of AIDS behavior among men in San Franscisco.* Subjects were recruited and surveyed annually. With complete data for 5 annual measurements, the goal of the analysis is to

---

* This description is based on a study conducted by McKusick, Coates, Morin, Pollack, and Hoff (1990). Neuhaus (1992) presents results for applying various panel data models.

determine the factors influencing an individual's probability of engaging in unsafe sexual behavior. The binary outcome is whether the individual engaged in unsafe behavior. The covariates include the age of the individual in years, an indicator of whether the man was involved in a primary gay relationship, an indicator of whether the man was involved in a monogamous relationship, an indicator of whether the man had been tested for HIV, and the number of AIDS symptoms reported for the year. For the following questions, indicate whether you would fit a population-averaged or a subject-specific model.

(a) Does the probability of engaging in unsafe behavior depend on the age of the individual?
(b) Does the unsafe behavior decrease over time for the population of men?
(c) Does the unsafe behavior decrease over time for an individual?
(d) Does the unsafe behavior decrease for an individual once they learn the results of their HIV test?
(e) Does the unsafe behavior decrease over time adjusted for the previous year?*

---

* This is a trick question!

CHAPTER 4

Residuals, Diagnostics, and Testing

This chapter highlights the techniques and measures used to evaluate GEE models. We also discuss techniques for choosing between competing models and the extensions of familiar diagnostics and graphical methods to GEE.

Each of the member distributions for the exponential family have implied quasilikelihoods. In Table 4.1 and Table 4.2, we list the log-likelihood functions and the implied quasilikelihood functions since they prove to be useful for several of the diagnostics presented in this chapter.

Family	$\mathcal{L}$
Gaussian	$-\frac{1}{2}\sum\left\{\frac{(y-\mu)^2}{\phi}+\ln(2\pi\phi)\right\}$
Binomial(k)	$\phi\sum\left\{\ln\Gamma(k+1)-\ln\Gamma(y+1)-\ln\Gamma(k-y+1)\right.$ $\left.+y\ln\left(\frac{\mu}{k}\right)+(\mu-y)\ln\left(1-\frac{\mu}{y}\right)\right\}$
Poisson	$\phi\sum\left\{-\mu+y\ln(\mu)-\ln\Gamma(y+1)\right\}$
Gamma	$-\frac{1}{\phi}\sum\left\{\frac{y}{\mu}-\ln\left(\frac{\phi}{\mu}\right)-\frac{\phi-1}{\phi}\ln(y)+\frac{1}{\phi}\ln\Gamma(\phi)\right\}$
Inverse Gaussian	$-\frac{1}{2}\sum\left\{\frac{(y-\mu)^2}{y\mu^2\phi}+\ln\left(\phi y^3\right)+\ln(2\pi)\right\}$

Table 4.1 *Log-likelihoods for exponential family members*

We first review the estimation procedure for the GEE1 for GLMs in order to establish needed quantities for defining the diagnostics required for model

Family	Q
Gaussian	$-\frac{1}{2}\sum(y-\mu)^2$
Binomial(k)	$\sum\{y\log(\mu/(1-\mu)) + \log(1-\mu)\}$
Poisson	$\sum\{y\log\mu - \mu\}$
Gamma	$-\sum\{y/\mu + \ln\mu\}$
Inverse Gaussian	$\sum\{-y/(2\mu^2) + 1/\mu\}$

Table 4.2 *Quasilikelihoods for exponential family members*

evaluation. Recall the estimating equation for the PA-GEE for GLMs:

$$\Psi(\boldsymbol{\beta},\boldsymbol{\alpha}) = (\Psi_{\boldsymbol{\beta}}(\boldsymbol{\beta},\boldsymbol{\alpha}), \Psi_{\boldsymbol{\alpha}}(\boldsymbol{\beta},\boldsymbol{\alpha})) \qquad (4.1)$$

$$= \begin{pmatrix} \sum_{i=1}^{n} \mathbf{x}_{ji}^{\mathrm{T}} \mathrm{D}\left(\frac{\partial \boldsymbol{\mu}_i}{\partial \boldsymbol{\eta}}\right) (\mathrm{V}(\boldsymbol{\mu}_i))^{-1} \left(\frac{\mathbf{y}_i - \boldsymbol{\mu}_i}{a(\phi)}\right) \\ \sum_{i=1}^{n} \left(\frac{\partial \boldsymbol{\xi}_i}{\partial \boldsymbol{\alpha}}\right)^{\mathrm{T}} \mathbf{H}_i^{-1} (\mathbf{W}_i - \boldsymbol{\xi}_i) \end{pmatrix} \qquad (4.2)$$

$$\mathrm{V}(\boldsymbol{\mu}_i) = \mathrm{D}(\mathrm{V}(\mu_{it}))^{1/2} \mathbf{R}(\boldsymbol{\alpha}) \mathrm{D}(\mathrm{V}(\mu_{it}))^{1/2} \qquad (4.3)$$

The heart of the procedure for solving this estimating equation is the iteratively reweighted least squares (IRLS) algorithm. This algorithm is a modification of the Newton–Raphson algorithm in which the expected Hessian matrix is substituted for the observed Hessian. The modification is known as the method of Fisher scoring. An updating equation for β is available under this approach such that

$$\widehat{\boldsymbol{\beta}}^{(r)} = \widehat{\boldsymbol{\beta}}^{(r-1)} - \left\{\sum_{i=1}^{n} \mathbf{D}_i^{\mathrm{T}} \mathrm{V}(\boldsymbol{\mu})_i^{-1} \mathbf{D}_i\right\} \left\{\sum_{i=1}^{n} \mathbf{D}_i^{\mathrm{T}} \mathrm{V}(\boldsymbol{\mu}_i)^{-1} \mathbf{S}_i\right\} \qquad (4.4)$$

where

$$\mathbf{D}_i = \mathrm{D}(\mathrm{V}(\mu_{it})) \mathrm{D}\left(\frac{\partial \boldsymbol{\mu}_i}{\partial \boldsymbol{\eta}}\right) \mathbf{X}_i \qquad (4.5)$$

$$\mathbf{S}_i = \mathbf{y}_i - g^{-1}(\widehat{\boldsymbol{\eta}}_i) \qquad (4.6)$$

In this form, the updating equation clearly has the form of a weighted least squares regression algorithm with a (synthetic) dependent variable given by the $(\sum n_i) \times 1$ column vector

$$\mathbf{Z}_{it} = (y_{it} - \widehat{\mu}_{it}) \left(\frac{\partial \mu}{\partial \eta}\right)_{it} + \widehat{\eta}_{it} \tag{4.7}$$

The weighted OLS algorithm involves covariates, $\mathbf{X}$, and weights

$$\mathbf{W}_i = \mathrm{D}\left\{\frac{1}{\mathrm{V}(\mu_{it})a(\phi)}\left(\frac{\partial \mu}{\partial \eta}\right)^2_{it}\right\} \tag{4.8}$$

Written explicitly, we see that the updating equation is

$$\beta^{\mathrm{new}} = (\mathbf{X}^\mathrm{T}\mathbf{W}\mathbf{X})^{-1}\mathbf{X}^\mathrm{T}\mathbf{W}\mathbf{Z} \tag{4.9}$$

The solution entails the alternating estimation of β and of $\boldsymbol{\alpha}$, with the results of each estimate being used to update values for the calculation of the subsequent estimate. Iterations continue until a predetermined criterion of convergence is reached. This is possible for the GEE1 models since we assume that the two estimating equations are orthogonal. While we have specifically emphasized the estimating equations for the PA-GEE models, the estimation steps are the same for alternating logistic regression (ALR) and the PA-EGEE models. However, the description given does not address estimation of GEE2 models.

4.1 Criterion measures

Several criterion measures have recently been proposed for evaluating GEE-constructed models. In the next few subsections we highlight several useful measures for evaluating the goodness of fit of the model, choosing the best correlation structure for a PA-GEE model, and choosing the best collection of covariates for a given correlation structure.

Akaike's information criterion (AIC) is a well-established goodness-of-fit statistic for likelihood-based model selection. Pan (2001a) introduced two useful extensions of this measure that we illustrate in the following subsections.

4.1.1 Choosing the best correlation structure

The AIC for likelihood-based models is defined as

$$\mathrm{AIC} = -2\mathcal{L} + 2p \tag{4.10}$$

where p is the number of parameters in the model. The goal is to generalize this measure for quasilikelihood models. Since $\mathcal{L}$, by definition, is the log-likelihood it seems obvious that we should be able to replace it with the quasilikelihood $\mathcal{Q}$. The penalty term in the AIC should also be generalized. Pan shows how we can derive a new measure called the *quasilikelihood under the independence model information criterion* (QIC).

Recall the quasilikelihood for PA-GEE models:

$$Q(y;\mu) = \int^{\mu} \frac{y-\mu^*}{V(\mu^*)} d\mu^* \tag{4.11}$$

We list quasilikelihoods for various distributions in the exponential family in Table 4.2. Regardless of the correlation structure $\mathbf{R}(\boldsymbol{\alpha})$ used in fitting the PA-GEE model, the quasilikelihood is calculated under the assumption of independence, $\mathbf{R} = \mathbf{I}$. It uses both the model coefficient estimates and the correlation in the process. However, the quasilikelihood does not itself directly address any type of correlation. The penalty term of the AIC, the $2p$ term, is calculated for the QIC as $2\,\mathrm{trace}\left(\mathbf{A}^{-1}\mathbf{V}_{\mathrm{MS},\mathbf{R}}\right)$ where $\mathbf{A}_\mathbf{I}$ is the variance matrix for the independence model and $\mathbf{V}_{\mathrm{MS},\mathbf{R}}$ is the sandwich estimate of variance for the correlated model. QIC($\mathbf{R}$) is defined from these terms as

$$\mathrm{QIC}(\mathbf{R}) = -2Q\!\left(g^{-1}(\mathbf{x}\boldsymbol{\beta}_\mathbf{R})\right) + 2\,\mathrm{trace}\left(\mathbf{A}_\mathbf{I}^{-1}\mathbf{V}_{\mathrm{MS},\mathbf{R}}\right) \tag{4.12}$$

The notation emphasizes:

- $Q\!\left(y; g^{-1}(\mathbf{x}\boldsymbol{\beta}_\mathbf{R})\right)$ is the value of the quasilikelihood computed using the coefficients from the model with hypothesized correlation structure $\mathbf{R}$. In evaluating the quasilikelihood, we use $\widehat{\boldsymbol{\mu}} = g^{-1}(\mathbf{x}\widehat{\boldsymbol{\beta}}_\mathbf{R})$ in place of $\boldsymbol{\mu}$ where $g^{-1}()$ is the inverse link function for the model.

- $\mathbf{A}_\mathbf{I}$ is the variance matrix obtained by fitting an independence model.

- $\mathbf{V}_{\mathrm{MS},\mathbf{R}}$ is the modified sandwich estimate of variance from the model with hypothesized correlation structure $\mathbf{R}$.

Since the definition of the QIC is in terms of the hypothesized correlation structure $\mathbf{R}$, we can use this measure to choose between several competing correlation structures. As with the AIC, the best model is the one with the *smallest* measure. The QIC is equal to the AIC when the model implies a likelihood proper and we are fitting an independence model (less a constant normalizing term).

We simulate data (see section 5.2.4) that follow an exchangeable correlation binomial-logit model where the common correlation is .4 for a balanced dataset with 50 individuals, each with 8 replicated observations. This simulation is similar to the one performed by Pan in the previously cited reference and is summarized in Table 4.3.

Using the QIC measure to choose among these 7 competing correlation structures leads us to select the exchangeable correlation model.

We also computed the QIC for several correlation structures using the Progabide data analyzed in Chapter 3. The results for various correlation structures are given in Table 4.4. Using the QIC measure to choose among the competing correlation structures illustrated in Chapter 3 again leads to the selection of the exchangeable correlation model.

In the previous chapter we also looked at simulated data (section 5.2.5) with a complicated correlation structure. We examined various methods for fitting an exchangeable correlation structure and, then in section 3.2.1.6, estimated a model matching the generating correlation structure. Computing the QIC

CRITERION MEASURES

Correlation	QIC
Exchangeable	449.7804
Independent	451.3903
AR(2)	451.7270
AR(1)	452.0540
Unstructured	452.2829
Nonstationary(2)	453.2632
Stationary(2)	453.4091

Table 4.3 *Simulation results for the QIC measure. The true correlation structure used in simulating the data is exchangeable.*

Correlation	QIC
Exchangeable	3206.677
AR(2)	3212.521
Unstructured	3225.236
Stationary(2)	3233.845
Nonstationary(2)	3233.845

Table 4.4 *QIC measures for several correlation structures for the PA-GEE Poisson model of the Progabide data.*

statistic for various correlation structures yields results in Table 4.5 validating the use of the more complicated correlation structure for fitting the model.

Correlation	QIC
Correct	171.894
AR(2)	172.917
Stationary(3)	173.390
Independent	173.656
Exchangeable	173.933

Table 4.5 *QIC measures for several correlation structures for the PA-GEE linear regression model of the data in section 5.2.5.*

In choosing the best correlation structure, we offer the following general guidelines.
- If the size of the panels is small and the data are complete, use the unstructured correlation specification.
- If the observations within a panel are collected for the same PSU over time, then use a specification that also has a time dependence.
- If the observations are clustered (not collected over time), then use the exchangeable correlation structure.

- If the number of panels is small, then the independence model may be the best; but calculate the sandwich estimate of variance for use with hypothesis tests and interpretation of coefficients.
- If more than one correlation specification satisfies the above descriptions, use the QIC measure to discern the best choice.

Of course, if there is motivating scientific evidence of a particular correlation structure, then that specification should be used. The QIC measure, like any model selection criterion, should not be blindly followed.

4.1.2 Choosing the best subset of covariates

The QIC_u is a measure that can be used to determine the best subset of covariates for a particular model. The measure is defined as

$$QIC_u = -2\mathcal{Q}(g^{-1}(\mathbf{x}\boldsymbol{\beta}_\mathbf{R})) + 2p \qquad (4.13)$$

where the notation emphasizes that the quasilikelihood is calculated for the independence model, but with the regression coefficients fitted for the hypothesized correlation structure.

In choosing between (two or more) models, the model with the smallest QIC_u criterion measure is preferred.

As a short first example, we look at the Progabide dataset. We fit all 1-factor and 2-factor models as well as the full (3-factor) models for comparison. The sorted results are presented in Table 4.6.

Covariates	QIC_u
time	3202.203
time progabide timeXprog	3206.677
time timeXprog	3207.649
time progabide	3209.472
progabide timeXprog	3253.112
progabide	3253.736
timeXprog	3257.007

Table 4.6 *QIC_u measures for models of the Progabide data.*

Using only the QIC_u measure, the best model includes only the time variable. Note that the difference in the criterion measure for the best model as well as for the full model is almost entirely due to the penalty $(2p)$ term. This criterion is meant as a guide for choosing between models when no scientific knowledge would guide the researcher to a preference. Despite the results of this investigation, we still prefer the full model.

4.2 Analysis of residuals

An analysis of data includes an important final check that the selected model adequately fits the data. This part of the analysis focuses on uncovering sig-

ANALYSIS OF RESIDUALS

nificant departures in the data from the model assumptions. We focus on two types of departure. The first is an observation (isolated) departure; the second is a model (systematic) departure.

4.2.1 A nonparametric test of the randomness of residuals

One can not apply many well-known techniques, without modification, to the case of PA-GEE for GLMs. Chang (2000) advises the use of the Wald–Wolfowitz run test to assist the analyst in uncovering possible patterns of nonrandomness using scatter plots of residuals. The test codes the residuals with an indicator of whether the residual is positive, ('1'), or negative, ('-1'). The sequence of codes is then examined and a count of the total number of runs of the two codes is computed. This is without regard to the length of any given run.

Let n_p indicate the total number of positive residuals, n_n indicate the total number of negative residuals, and T indicate the number of observed runs in our sequence. Under the null hypothesis that the signs of the residuals are distributed in a random sequence, the expected value and variance of T are:

$$\mathrm{E}(T) = \frac{2n_p n_n}{n_p + n_n} + 1 \tag{4.14}$$

$$\mathrm{V}(T) = \frac{2n_p n_n (2n_p n_n - n_p - n_n)}{(n_p + n_n)^2 (n_p + n_n - 1)} \tag{4.15}$$

A test statistic for this hypothesis is then

$$W_Z = \frac{T - \mathrm{E}(T)}{\sqrt{\mathrm{V}(T)}} \tag{4.16}$$

which has an approximately standard normal distribution. Extreme values of W_Z indicate that the model does not adequately reflect the underlying structure of the data.

Clearly, this test relies on a specific ordering of the residuals. As such, the test may be amended in order to assess different hypotheses. An overall test of the panel structure of the model could sort the residuals in the "natural order." That is, the data would be sorted by the panel identifier i and the repeated measures identifier t within i. Alternatively, if we wish to assess whether a given (continuous) covariate is specified in the correct functional form, we can sort the residuals on that covariate, or we can test the model adequacy sorting on the fitted values.

4.2.2 Graphical assessment

The first step in an exploratory data analysis (EDA) should include a graphical illustration of the raw data. To accomplish this, we want to include illustrations of the data that reflect the panel nature. One such approach (for nonbinomial models) shows boxplots of the outcome for each of the repeated measures. Using the Progabide data, we can illustrate boxplots for the baseline and four follow-ups of the seizure counts.

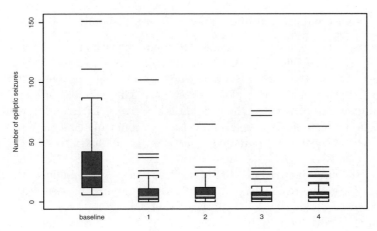

Boxplots of seizures by observation time

Since we use the log link with Poisson variance to model the counts, we can also illustrate the log of the seizure counts.

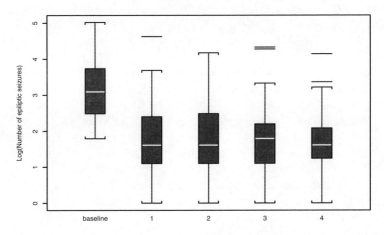

Boxplots of log(seizures) by observation time

These boxplots show the raw and log transformed seizure counts for the entire dataset. Our analysis is primarily focused on the efficacy of the Progabide treatment. As such, we also illustrate boxplots of placebo and Progabide observations for each observation time. The log transformed data are used for illustration.

Boxplots for the placebo are

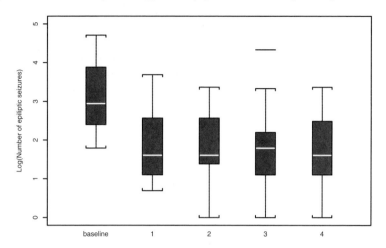

and boxplots for the Progabide treatment are displayed as

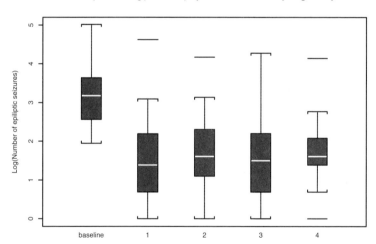

Standard approaches used for model building apply equally for GEE models. One should assess model adequacy via the same types of residual plots used in linear models with the added requirement that the illustration should identify the panel structure of the data.

In the previous chapter we used the two simulated datasets from section 5.2.5 (linear data) and section 5.2.4 (logistic data). Here we illustrate the resulting residual plots for fitting a number of different assumed correlation structures in a PA-GEE model. The data are generated from a linear model with a correlation structure not (directly) supported by software. The logistic data are generated from a logistic model where the data are characterized by an exchangeable correlation structure.

Suppose we wish to examine the results for fitting an independent correlation, an exchangeable correlation, and the theoretically correct correlation structure to the linear data. The results of fitting the independent correlation model to linear data are:

```
GEE population-averaged model              Number of obs       =          80
Group variable:                   id       Number of groups    =          10
Link:                       identity       Obs per group: min  =           8
Family:                     Gaussian                      avg  =         8.0
Correlation:             independent                      max  =           8
                                           Wald chi2(2)        =       50.54
Scale parameter:            1.029406       Prob > chi2         =      0.0000

Pearson chi2(80):              82.35       Deviance            =       82.35
Dispersion (Pearson):       1.029406       Dispersion          =    1.029406

------------------------------------------------------------------------------
           y |      Coef.   Std. Err.       z    P>|z|     [95% Conf. Interval]
-------------+----------------------------------------------------------------
          x1 |   1.158479   .2397294      4.83   0.000     .6886185    1.62834
          x2 |   1.158928   .2708094      4.28   0.000     .6281517   1.689705
       _cons |   .8813407    .238173      3.70   0.000     .4145301   1.348151
------------------------------------------------------------------------------
```

The residual plot for this model is illustrated below.

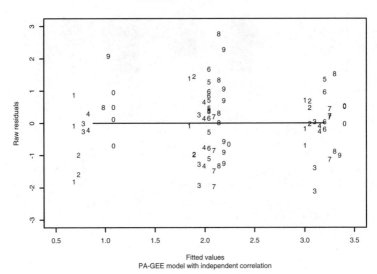

Residuals versus Fitted Values

Fitted values
PA-GEE model with independent correlation

ANALYSIS OF RESIDUALS

Since there are only 3 distinct covariate patterns in these data, the residual plot includes only 3 distinct values on the horizontal axis. We moved each identifier slightly (in the horizontal direction) to more clearly observe the panel identifiers of the residuals; this is a standard graphical technique called jitter. Plots of this type are routinely examined to see if residuals in each panel have the same sign.

We calculate the runs test to examine the randomness of the residuals. Below is a graphical illustration of the test.

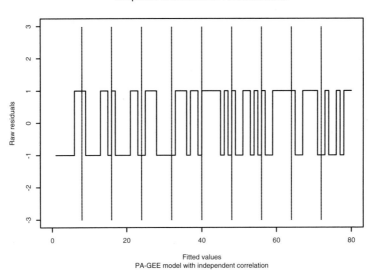

Graphical Illustration of Residual Runs

Test results provide the following statistics:

$$n_p = 42 \qquad (4.17)$$
$$n_n = 38 \qquad (4.18)$$
$$T = 44 \qquad (4.19)$$
$$E(T) = \frac{2n_p n_n}{n_p + n_n} + 1 = 40.9 \qquad (4.20)$$
$$V(T) = \frac{2n_p n_n (2n_p n_n - n_p - n_n)}{(n_p + n_n)^2 (n_p + n_n - 1)} = 19.65 \qquad (4.21)$$
$$Z = \frac{44 - 40.9}{\sqrt{19.65}} = 0.6993 \qquad (4.22)$$
$$p = .2422 \qquad (4.23)$$

The test reveals that there is not enough evidence to reject the hypothesis that the residuals from the model are random. In general, the result of the runs test does not significantly change due to the hypothesized structure when the model is correct in terms of including necessary covariates in their proper form. We instead use the QIC measure to select the best correlation structure instead.

The graphical illustration of the residual runs test is produced by plotting $\text{sign}(\hat{r}_{it})$ versus the observation number (where the data are sorted by panel identifier and by repeated measure identifier within the panel number). The vertical lines indicate the number of runs in the residuals (less one). The grid lines indicate the breaks in the panels and allow us another method for checking the number of panels where the residuals have a common sign. Preferably, we would still produce this plot for a slightly larger dataset, but break the presentation into several smaller units. This type of plot is not useful for very large datasets since the amount of information becomes too dense. Hence, the plot is rather indecipherable.

For the logistic data, we can first fit an independence model:

```
GEE population-averaged model            Number of obs      =       400
Group variable:                    id    Number of groups   =        50
Link:                           logit    Obs per group: min =         8
Family:                      binomial                   avg =       8.0
Correlation:             exchangeable                   max =         8
                                         Wald chi2(3)       =     20.48
Scale parameter:                    1    Prob > chi2        =    0.0001

------------------------------------------------------------------------
       y |      Coef.   Std. Err.      z    P>|z|    [95% Conf. Interval]
---------+--------------------------------------------------------------
      x1 |  -.3325008   .3533269    -0.94   0.347   -1.025009    .3600073
      x2 |   .2515805   .1081765     2.33   0.020    .0395584    .4636025
      x3 |   .1423381   .0374987     3.80   0.000     .068842    .2158342
   _cons |   1.277268   .2843566     4.49   0.000    .7199388    1.834596
------------------------------------------------------------------------
```

We plot the Pearson residuals for all panels for each of the 8 repeated values in order to examine whether there is an order effect.

ANALYSIS OF RESIDUALS

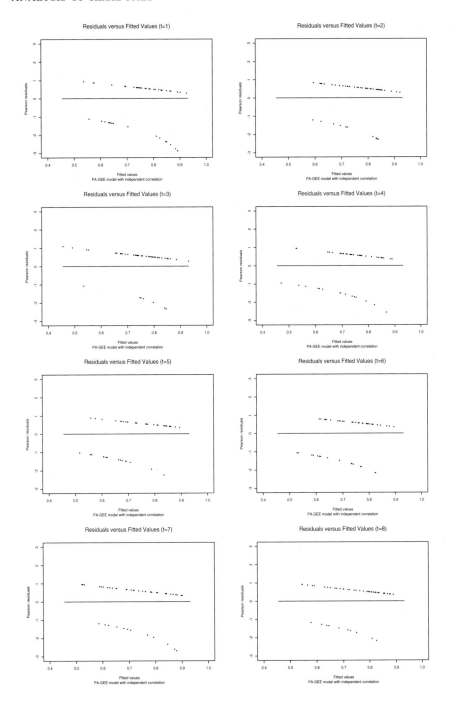

There is no indication in the plots that the residuals depend on either the panel identifier or on the repeated measures identifier. In this case, all of the plots are similar. As previously discussed, we can also test the residuals. A graphical illustration of the runs test displays as:

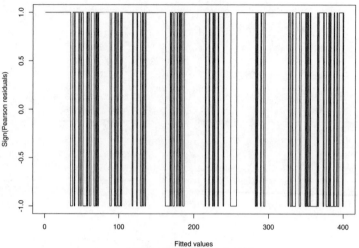

The test results are as follows:

$$n_p = 296 \tag{4.24}$$
$$n_n = 104 \tag{4.25}$$
$$T = 291 \tag{4.26}$$
$$\mathrm{E}(T) = \frac{2n_p n_n}{n_p + n_n} + 1 = 154.92 \tag{4.27}$$
$$\mathrm{V}(T) = \frac{2n_p n_n (2n_p n_n - n_p - n_n)}{(n_p + n_n)^2 (n_p + n_n - 1)} = 58.99 \tag{4.28}$$
$$Z = \frac{291 - 154.92}{\sqrt{58.99}} = 17.72 \tag{4.29}$$
$$p < .0001 \tag{4.30}$$

In this case, the test provides a clear indication that the residuals are not random. In fact, we know that these particular data are generated from a correlation structure not directly supported by standard commercial software.

The runs test is useful for examining the dependence of the residuals on the individual covariates. For this example, we generated data that accord with the model

$$y_{it} = 1 + \text{x1} + 2\text{x1}^2 + \nu_i + \epsilon_{it} \tag{4.31}$$

where $\nu \sim \mathrm{N}(0, \sigma_\nu^2)$ and $\epsilon \sim \mathrm{N}(0, 1)$. This is an exchangeable correlation model where $\rho = .3$, the correlation used in generating the data.

Suppose that we misspecify the model in the estimation using only the x1 variable without modeling the square of the covariate x12. Two techniques ex-

ANALYSIS OF RESIDUALS

ist for discovering this missing covariate. The first method is graphical where we plot the residuals versus the covariate; the second method used is to calculate the runs test where the data are sorted according to the covariate of interest. This is possible when the covariate under investigation is continuous.

The fit of the independence model using both covariates is presented as:

```
GEE population-averaged model           Number of obs      =       100
Group variable:                    id   Number of groups   =        25
Link:                        identity   Obs per group: min =         4
Family:                      Gaussian                  avg =       4.0
Correlation:              independent                  max =         4
                                        Wald chi2(2)       =  73374.59
Scale parameter:             1.063566   Prob > chi2        =    0.0000

Pearson chi2(100):             106.36   Deviance           =    106.36
Dispersion (Pearson):        1.063566   Dispersion         =  1.063566

------------------------------------------------------------------------
     y |     Coef.   Std. Err.      z    P>|z|    [95% Conf. Interval]
-------+----------------------------------------------------------------
    x1 |  .9963109   .0371903    26.79   0.000    .9234193    1.069202
   x12 |  2.006862    .007409   270.87   0.000    1.99234     2.021383
  _cons|  .8291631   .1187399     6.98   0.000    .5964372    1.061889
------------------------------------------------------------------------
```

The fit of the independence model, excluding the x12 variable, is:

```
GEE population-averaged model           Number of obs      =       100
Group variable:                    id   Number of groups   =        25
Link:                        identity   Obs per group: min =         4
Family:                      Gaussian                  avg =       4.0
Correlation:              independent                  max =         4
                                        Wald chi2(1)       =      0.01
Scale parameter:             781.3898   Prob > chi2        =    0.9294

Pearson chi2(100):           78138.98   Deviance           =  78138.98
Dispersion (Pearson):        781.3898   Dispersion         =  781.3898

------------------------------------------------------------------------
     y |     Coef.   Std. Err.      z    P>|z|    [95% Conf. Interval]
-------+----------------------------------------------------------------
    x1 |  .0889191   1.003949     0.09   0.929   -1.878785    2.056623
  _cons|   16.6458   2.802395     5.94   0.000   11.15321    22.13839
------------------------------------------------------------------------
```

We calculate the runs test for the above model (where the data are in natural order) to obtain the following results:

$$n_p = 22 \qquad (4.32)$$
$$n_n = 78 \qquad (4.33)$$
$$T = 37 \qquad (4.34)$$
$$E(T) = \frac{2 n_p n_n}{n_p + n_n} + 1 = 36.32 \qquad (4.35)$$
$$V(T) = \frac{2 n_p n_n (2 n_p n_n - n_p - n_n)}{(n_p + n_n)^2 (n_p + n_n - 1)} = 11.55 \qquad (4.36)$$
$$Z = \frac{37 - 36.32}{\sqrt{11.55}} = 0.200 \qquad (4.37)$$
$$p = .4207 \qquad (4.38)$$

The test does not indicate nonrandomness in the residuals, when the residuals are ordered first by panel identifier and then by the repeated measures identifier within the panel identifier. A plot of the residuals sorted in this manner supports the conclusions of the test, while still indicating a misspecification of indeterminate nature.

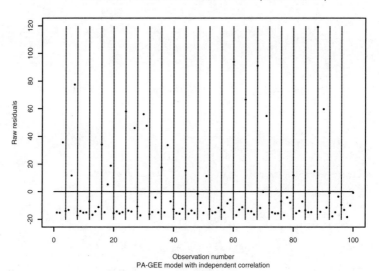

Residuals versus Observation Number (Natural Order)

Observation number
PA-GEE model with independent correlation

The plot clearly shows that the magnitude of the positive residuals is much larger than the magnitude of the negative residuals. However, there is no indication of the causal source. The vertical lines in the plot distinguish the individual panels, where the residuals do not indicate a time dependence.

The plot of the raw residuals versus the fitted values is given below.

ANALYSIS OF RESIDUALS

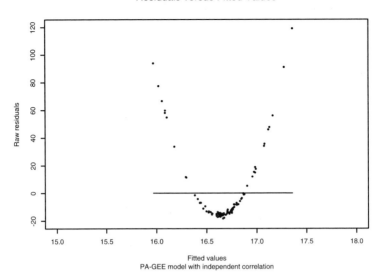

A plot of the raw residuals versus the x1 covariate values is displayed as:

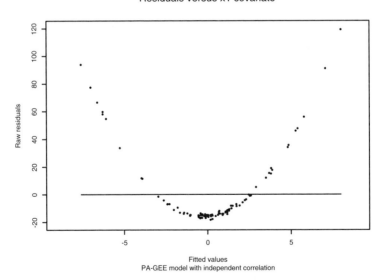

The two previous plots clearly indicate model misspecification. Similarly, the graphical illustration of the runs test shows a deviation from randomness for the residuals when the residuals are sorted for the values of the x1 covariate.

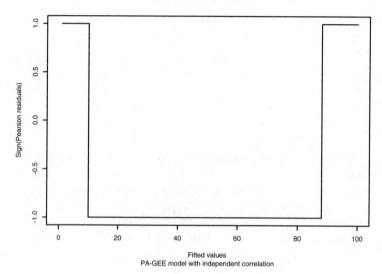

Graphical Illustration of Residual Runs

Calculating the runs test for this arrangement of the residuals results in

$$n_p = 22 \tag{4.39}$$
$$n_n = 78 \tag{4.40}$$
$$T = 3 \tag{4.41}$$
$$E(T) = \frac{2n_p n_n}{n_p + n_n} + 1 = 35.32 \tag{4.42}$$
$$V(T) = \frac{2n_p n_n (2n_p n_n - n_p - n_n)}{(n_p + n_n)^2 (n_p + n_n - 1)} = 11.55 \tag{4.43}$$
$$Z = \frac{3 - 35.32}{\sqrt{11.55}} = -9.510 \tag{4.44}$$
$$p < .0001 \tag{4.45}$$

Test results coincide with the graphical assessment. We have strong evidence that the covariate x1 is misspecified in the fitted model. For a continuous outcome model, we can plot the residuals versus each of the sorted covariates.

In addition to the graphical assessment and the nonparametric run test, we can also calculate the QIC_u criterion measure for the competing models.

$$QIC_u(\texttt{x1}) = 156354.7721 \tag{4.46}$$
$$QIC_u(\texttt{x1 x12}) = 220.222 \tag{4.47}$$

The results agree with our other model analyses.

4.2.3 Quasivariance functions for PA-GEE models

Wedderburn (1974) includes an analysis of Sitka spruce data. The author uses these data to illustrate that the usual binomial variance function does

ANALYSIS OF RESIDUALS

not adequately model the variance of the data. The data used in the analysis are actually at the individual level, but include an identifier for the variety of the barley. We use the variety as the panel identifier and fit a marginal model where we hypothesize that the observations within variety share a common correlation.

The data include the percentage `leaf` of the leaf area of barley affected by *Rhynchosporium secalis*, or leaf blotch, and binary variables `site1`,... ,`site9` to indicate the site at which the data are collected. The response variable `leaf` was set to .01% for those observations that were zero (as in the original analysis). There are 10 panels in the data and 9 sites for a total of 90 observations.

This analysis will require specification of a variance function that is not part of the usual collection of functions defined from the members of the exponential family. The quasivariance functions that we specify must be programmed. For this analysis, the S-PLUS package provides the best support for user-written variance functions. We utilize standard plots to assess the adequacy of the model.

The fit of an exchangeable logistic PA-GEE model provides

```
Coefficients:
                Values    Stderr  t-values Pr(|t|>)
(Intercept)  0.4683789 1.2548246  0.3732625   0.7099
      site1 -5.7024161 0.8578567 -6.6472829   0.0000
      site2 -4.1052059 1.3465430 -3.0487002   0.0031
      site3 -2.5554481 1.0323553 -2.4753572   0.0154
      site4 -2.3385667 1.3778087 -1.6973087   0.0935
      site5 -2.3378754 0.8762535 -2.6680355   0.0092
      site6 -2.0826245 1.4303881 -1.4559857   0.1493
      site7 -1.4372780 1.2446647 -1.1547511   0.2516
      site8 -0.8738440 0.4381757 -1.9942776   0.0495

Degrees of Freedom: 90 Total; 81 Residual
> summary(res)
      EFFECTS NDF DDF       F P.value
1 (Intercept)   1  81  0.1393  0.7100
2       site1   1  81 44.1864  0.0000
3       site2   1  81  9.2946  0.0031
4       site3   1  81  6.1274  0.0154
5       site4   1  81  2.8809  0.0935
6       site5   1  81  7.1184  0.0092
7       site6   1  81  2.1199  0.1493
8       site7   1  81  1.3335  0.2516
9       site8   1  81  3.9771  0.0495
```

A plot of the Pearson residuals versus the linear predictor is given by

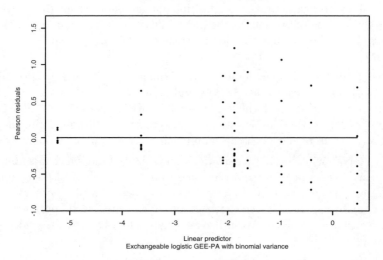

and a plot of the Pearson residuals versus the log of the variance for the model is

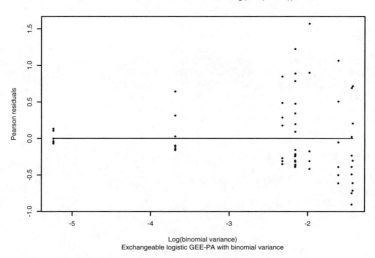

The plots indicate a lack of fit for very large and very small response outcomes. The plot of the Pearson residuals versus the log of the variance should look more uniform if the model were truly adequate for the data.

Following the analysis of Wedderburn, we hypothesize a quasivariance function that is the square of the usual binomial variance. We need the ability to specify a variance function to fit such a model. Most current software packages do not allow this specification. Among the packages that are used throughout this text, S-PLUS does support this specification.

Programming the squared binomial variance function into the software en-

ANALYSIS OF RESIDUALS

tails the additional specification of a deviance function. The fit of an exchangeable logistic PA-GEE model with quasi (square binomial) variance provides the following output:

```
Coefficients:
              Values    Stderr   t-values Pr(|t|>)
(Intercept)  0.4683789 1.2548246  0.3732625  0.7099
      site1 -5.7126423 0.8601273 -6.6416246  0.0000
      site2 -4.1056059 1.3466953 -3.0486524  0.0031
      site3 -2.5555499 1.0325136 -2.4750764  0.0154
      site4 -2.3385667 1.3778087 -1.6973087  0.0935
      site5 -2.3378754 0.8762535 -2.6680355  0.0092
      site6 -2.0826245 1.4303881 -1.4559857  0.1493
      site7 -1.4372780 1.2446647 -1.1547511  0.2516
      site8 -0.8738440 0.4381757 -1.9942776  0.0495

Degrees of Freedom: 90 Total; 81 Residual
> summary(res)
       EFFECTS NDF DDF        F  P.value
1 (Intercept)   1  81   0.1393   0.7100
2        site1  1  81  44.1112   0.0000
3        site2  1  81   9.2943   0.0031
4        site3  1  81   6.1260   0.0154
5        site4  1  81   2.8809   0.0935
6        site5  1  81   7.1184   0.0092
7        site6  1  81   2.1199   0.1493
8        site7  1  81   1.3335   0.2516
9        site8  1  81   3.9771   0.0495
```

A plot of the Pearson residuals versus the linear predictor is displayed as

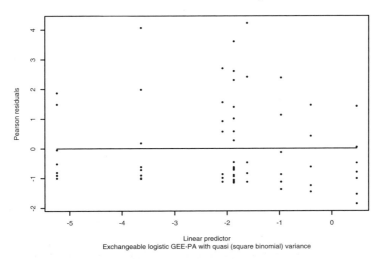

and a plot of the Pearson residuals versus the log of the variance for the model yields

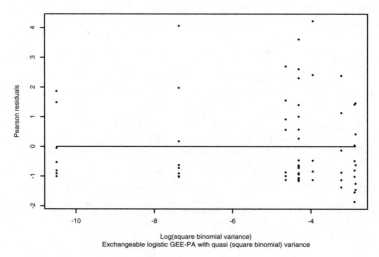

Pearson residuals versus log(mu^2(1-mu)^2)

Log(square binomial variance)
Exchangeable logistic GEE-PA with quasi (square binomial) variance

As in Wedderburn's original analysis, we see a substantial reduction for the effect of the extreme fitted values. The model has an overall better fit for the data.

The original analysis used the panels as another collection of fitted values; the analysis was a GLM. A true analysis of these data would allow the dispersion parameter to freely vary. There is no *a priori* reason to assume the standard binomial value of 1 for the dispersion when the outcomes represent percentages rather than binary outcomes. We also emphasize that our focus in this example is on the specification of alternate quasivariance functions. In so doing, we adopted one of the original covariates in the designed experiment to serve as our panel identifier. While the specification of the quasivariance shows improvement over the specification of the binomial variance, the panel analysis is not preferred over Wedderburn's original illustrations.

4.3 Deletion diagnostics

GLM analysis utilizes the DFBETA and DFFIT residuals described for general OLS regression in Belsey, Kuh, and Welsch (1980), and Cook's distance for identifying isolated departures. To check model departures, one usually relies on scatter plots of both raw and standardized residuals versus fitted values (as we demonstrated) as well as on other prognostic factors.

In this section we address the methods of case deletion. This is a well known diagnostic tool used extensively in OLS regression. In the usual GLM analysis, one may refit models leaving out each observation in turn to assess the impact of the change in the fitted model for a given observation. We note here that Stata's `glm` command, created by the authors, supports several deletion, jackknife, and bootstrapping techniques as command options. These techniques extend to GEE models as well.

In the case of the GEE1 for GLMs, we must address the panel structure of the data as Haslett (1999) did for the linear model with correlated errors. Preisser and Qaqish (1996) consider diagnostics for these models that measure the influence of a subset of observations either on the estimated regression parameters or on the linear predicted values. More importantly, the authors provide a simple formula for one-step estimates of the measures of influence.

These diagnostics provide the data analyst with the tools to identify those panels, or individual observations, having an undue influence on the fitted model. We could refit the model with the associated subset of observations deleted in order to obtain the exact measure of influence. In fact, we could do this for every subset. However, this becomes more of a time constraint, since as our datasets grow, it becomes desirable and at times even necessary to develop one-step approximations to the influence.

The basic idea for one-step approximations is to restart the estimation using the full sample estimates of β and α. We delete the associated subset of observations, and then reestimate the two parameter vectors with only one iteration of the estimation procedure.

Deletion diagnostics provide a measure of the influence of observations on parameter estimates and fitted values. When only one observation is left out in a deletion diagnostic procedure, it is called an *observation-deletion* diagnostic. It is called a *cluster-deletion* diagnostic when a set of observations corresponds to a cluster or panel.

As previously described, there are two approaches to the construction of influence measures. In one approach, we can measure the difference in the fitted regression coefficients by deleting a single observation from the estimation. In the second approach, we measure the difference in the fitted regression coefficients by deleting a panel of observations.

Let k index the subset of k observations that are to be deleted, and let $[k]$ denote the remaining observations. It follows then that $\widehat{\beta}_{[k]}$ denotes the estimated regression parameters with the set of k observations that are deleted. Lastly, we provide the equations $\mathbf{Q} = \mathbf{X}(\mathbf{X}^T\mathbf{W}\mathbf{X})^{-1}\mathbf{X}^T$ and $\mathbf{H} = \mathbf{Q}\mathbf{W}$. These equations serve as a basis for the discussion to follow in the next two subsections.

4.3.1 Influence measures

The DFBETA diagnostic is a measure of the difference between the full sample estimator $\widehat{\beta}$ and the fitted coefficient vector based on deleting one or more observations. A one-step approximation for the difference of the full sample coefficient vector and the estimated coefficient vector where an entire panel of observations is deleted is

$$(\mathbf{X}^T\mathbf{W}\mathbf{X})^{-1}\mathbf{x}_i^T(\mathbf{w}_i^{-1} - \mathbf{h}_i)^{-1}\mathbf{S}_i \qquad (4.48)$$

A one-step approximation for the difference of the full sample coefficient vector and the estimated coefficient vector where a single observation is deleted is

$$(\mathbf{X}^{\mathrm{T}}\mathbf{W}\mathbf{X})^{-1}\mathbf{x}_{it}^{\mathrm{T}}\frac{S_{it}}{\mathbf{w}_{it} - \mathbf{h}_{it}} \tag{4.49}$$

One may either use the one-step approximations to the estimated coefficients for leaving out observations, or one may fully fit the model for each subsample.

We have looked several times at the Progabide data. Here, the calculation of diagnostic measures for the covariates is examined. A Poisson model is used to explain the number of seizures experienced by patients in the panel study. The outcome of the analysis is

```
GEE population-averaged model           Number of obs       =      295
Group variable:                    id   Number of groups    =       59
Link:                             log   Obs per group: min  =        5
Family:                       Poisson                  avg  =      5.0
Correlation:              exchangeable                 max  =        5
                                        Wald chi2(3)        =    13.73
Scale parameter:                    1   Prob > chi2         =   0.0033

------------------------------------------------------------------------
    seizures |    Coef.   Std. Err.      z    P>|z|   [95% Conf. Interval]
-------------+----------------------------------------------------------
        time |  .111836   .0347333    3.22   0.001    .0437601    .179912
    progabide|  .0275345  .0466847    0.59   0.555   -.0639658   .1190348
   timeXprog | -.1047258  .0494024   -2.12   0.034   -.2015526  -.0078989
       _cons |  1.347609  .0340601   39.57   0.000    1.280853   1.414366
    lnPeriod |  (offset)
------------------------------------------------------------------------
```

The model uses a baseline measure for the offset and includes covariates on time, an indication of Progabide treatment, and a time by treatment interaction. The particular PA-GEE model fit here is an exchangeable correlation Poisson.

The dataset is relatively small, so that the DFBETA statistics are calculated using both the one-step approximations and by refitting the model to the data subsets. In addition, the DFBETA measures are examined for both the individual observation and the panel.

We next present a plot of the panel-level DFBETA statistics for the time covariate where the DFBETA statistics are calculated by refitting the model:

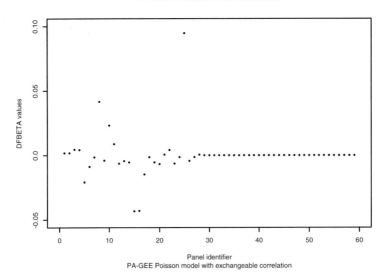

Next is a plot of the panel-level DFBETA statistics for the `progabide` covariate, where the DFBETA statistics are calculated by refitting the model:

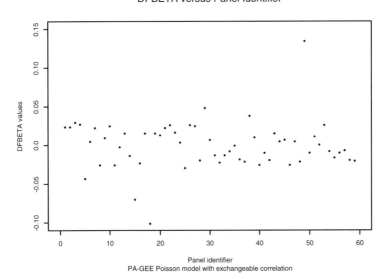

and now a plot of the panel-level DFBETA statistics for the `time-progabide` interaction where the DFBETA statistics are calculated by refitting the model:

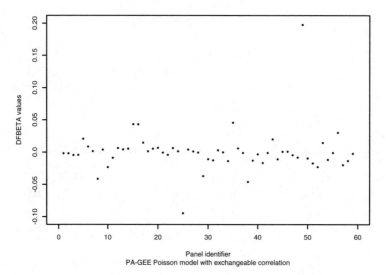

DFBETA versus Panel Identifier
PA-GEE Poisson model with exchangeable correlation

This collection of plots investigates the outliers of the covariates and their effect on the fitted model. There is evidence that the patient identified by panel id = 49 has an unusually large effect on two of the covariates in the analysis. We should point out that the data listed in this text have no other patient identifiers; the same data appear in other sources where this patient is identified as "patient 207"; see Diggle, Liang, and Zeger (1994).

Knowing that this patient seems to have a large effect on the outcomes, we could proceed with an exploratory analysis without this particular patient.

```
GEE population-averaged model              Number of obs      =         290
Group variable:                  id        Number of groups   =          58
Link:                           log        Obs per group: min =           5
Family:                     Poisson                       avg =         5.0
Correlation:           exchangeable                       max =           5
                                           Wald chi2(3)       =       27.56
Scale parameter:                  1        Prob > chi2        =      0.0000

------------------------------------------------------------------------------
    seizures |     Coef.   Std. Err.      z    P>|z|     [95% Conf. Interval]
-------------+----------------------------------------------------------------
        time |   .111836   .0378762     2.95   0.003     .0376001    .186072
    progabide|  -.1068224  .0486304    -2.20   0.028    -.2021362   -.0115087
   timeXprog |  -.3023841  .0595185    -5.08   0.000    -.4190382    -.18573
       _cons |   1.347609  .0340601    39.57   0.000     1.280853   1.414366
    lnPeriod |   (offset)
------------------------------------------------------------------------------
```

While there seems to be evidence that this patient is different from the rest, without investigating the causes, we should not drop the patient from the analysis simply on the basis of the results of DFBETA investigation. The

DELETION DIAGNOSTICS

patient does seem to have an extraordinarily high number of seizures in the baseline, perhaps indicating other medical conditions. In this case, we would contact the collectors of the data for further explanation.

Continuing our investigation, we can also calculate DFBETA measures on an observation basis rather than a panel basis. Here is a plot of the observation-level DFBETA statistics for the `time` covariate where the DFBETA statistics are calculated by refitting the model:

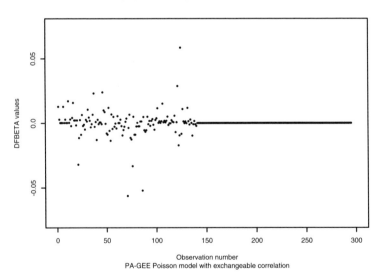

We now present a plot of the observation-level DFBETA statistics for the `progabide` covariate where the DFBETA statistics are calculated by refitting the model:

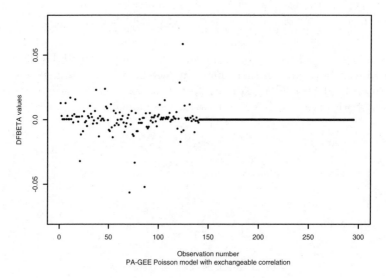

Here is a plot of the observation-level DFBETA statistics for the `time-progabide` interaction where the DFBETA statistics are calculated by refitting the model:

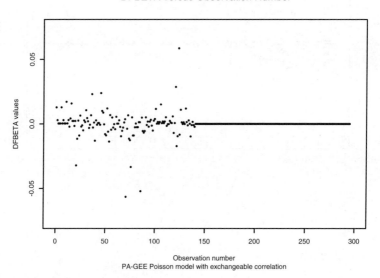

We may have anticipated that the analysis of the DFBETA statistics for the observations would coincide with the analysis for the DFBETA statistics calculated for the panels. However, this is not the case. When we delete only single observations, the panel effect, for which the particular observation is a member, is not removed unless the observation represents a (singleton) panel.

4.3.2 Leverage measures

We may also extend Cook's distance for use with panel data in addition to measuring the influence of observations on the estimated coefficient vector. This is a standardized measure of the influence of a set of observations on the linear predicted value. DFBETA residuals are used to investigate the effect of outliers in the covariates; Cook's distance is used to investigate the effect of the outliers in the outcome.

Cook's distance, structured as leaving out an entire panel, is defined as

$$\text{Cook}_i = \left(\frac{1}{p\widehat{\phi}}\right) \mathbf{S}_i^\text{T}(\mathbf{W}_i^{-1} - \mathbf{Q}_i)^{-1}\mathbf{Q}_i(\mathbf{W}_i^{-1} - \mathbf{Q}_i)^{-1}\mathbf{S}_i \qquad (4.50)$$

and for the case of leaving out a single observation, Cook's distance is defined as

$$\text{Cook}_{it} = \frac{\mathbf{S}_{it}^2 \mathbf{Q}_{it}}{p\widehat{\phi}(\mathbf{W}_{it}^{-1} - \mathbf{Q}_{it})^2} \qquad (4.51)$$

Cook's distance is a scaled measure of the distance between the coefficient vectors when the kth group of observations is deleted from the analysis.

We should also consider methods for calculating the leverage of the observations on the estimated association vector when the association parameters are of prime interest. This type of analysis also necessitates the need to estimate the variance of these parameters.

4.4 Goodness of fit (population-averaged models)

Zheng (2000) provides discussion of measures of goodness of fit for PA-GEE for GLMs. These measures are generalizations of measures commonly used for assessing the goodness of fit of GLMs.

4.4.1 Proportional reduction in variation

Similar to a GLM, the marginal model of a PA-GEE specifies a conditional mean and a link function. The variance of the response and the block-diagonal covariance matrix describing the intrapanel correlation among the repeated responses are functions of the mean and possible additional parameters $\boldsymbol{\alpha}$.

Since the marginal model does not specify a likelihood, we must consider extensions of nonlikelihood based summary measures in order to analyze the goodness of fit. The extension of the entropy H measure for categorical marginal models is defined as

$$H_{\text{MARG}} = 1 - \frac{\sum_{i=1}^{n}\sum_{t=1}^{n_i}\sum_{k=1}^{K}\widehat{\pi}_{itk}\ln(\widehat{\pi}_{itk})}{N\sum_{k=1}^{K}\widehat{\alpha}_k\ln(\widehat{\alpha}_k)} \qquad (4.52)$$

The interpretation of H_{MARG} can be thought of as the proportional reduction in entropy due to the model of interest. It is equal to the usual measure for GLM when the number of observations per panel is one ($n_i = 1$ for all i).

An extension of the R^2 measure is calculated using

$$R^2_{\text{MARG}} = 1 - \frac{\sum_{i=1}^{n} \sum_{t=1}^{n_i} \left(Y_{it} - \widehat{Y}_{it}\right)^2}{\sum_{i=1}^{n} \sum_{t=1}^{n_i} \left(Y_{it} - \overline{Y}\right)^2} \qquad (4.53)$$

This measure is interpreted as the proportion of variance in the outcome that is explained by the model.

4.4.2 Concordance correlation

The concordance correlation is a statistic of the agreement of two persons or measures. The observed and fitted values from a given model are the two measures for which we calculate the correlation.

$$r_c = \frac{2 \sum_{i=1}^{n} \sum_{t=1}^{n_i} (Y_{it} - \overline{Y}_{..})(\widehat{Y}_{it} - \overline{\widehat{Y}}_{..})}{\sum_{i=1}^{n} \sum_{t=1}^{n_i} (Y_{it} - \overline{Y}_{..})^2 + \sum_{i=1}^{n} \sum_{t=1}^{n_i} (\widehat{Y}_{it} - \overline{\widehat{Y}}_{..})^2} \qquad (4.54)$$

This measure is less than or equal to Pearson's correlation coefficient. The reason for this is that the concordance correlation imposes the constraint that the best fitting line goes through the origin with slope 1 when comparing the observed and fitted values.

Here we investigate the concordance correlation coefficient utilizing calculations from Bland and Altman (1986) and Lin (1989). The coefficient is calculated for the fit of a marginal model for the Progabide data with the observed values. This investigation produces the following results

```
Concordance correlation coefficient (Lin, 1989)

 rho_c   SE(rho_c)   Obs    [   95% CI   ]     P        CI type
-----------------------------------------------------------------
 0.391     0.037     295    0.318  0.463    0.000     asymptotic
                            0.316  0.461    0.000     z-transform

Pearson's r =  0.493   Pr(r = 0) = 0.000   C_b = rho_c/r =  0.793
Reduced major axis:    Slope =     2.029   Intercept  =  -13.243

   Difference (shat - seizures)        95% Limits Of Agreement
    Average       Std. Dev.             (Bland & Altman, 1986)
-----------------------------------------------------------------
     0.000         16.262               -31.872        31.872
```

We can graphically depict the concordance by plotting the observed values by the fitted values. In this plot, we include the lines of perfect fit for the Pearson and concordance correlations.

GOODNESS OF FIT (POPULATION-AVERAGED MODELS) 167

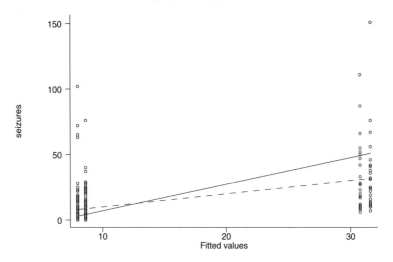

4.4.3 A χ^2 goodness of fit test for PA-GEE binomial models

Horton, Bebchuk, Jones, Lipsitz, Catalano, Zahner, and Fitzmaurice (1999) present an extension of the Hosmer Jr. and Lemeshow (1980) goodness of fit test applicable to PA-GEE models. The basic idea of the test is to group the ordered fitted values into groups defined by deciles (tenth percentile, twentieth percentile, etc.), and calculate a χ^2 goodness of fit test on the counts.

The ordered fitted values $\widehat{\mu}_{it}$ are used to define groups defined by

1. The first group contains the $\sum_i n_i/10$ observations associated with the smallest $\widehat{\mu}_{it}$ fitted values.

2. The second group contains the $\sum_i n_i/10$ observations associated with the next smallest $\widehat{\mu}_{it}$ fitted values.

$\vdots$

10. The tenth (last) group contains the $\sum_i n_i/10$ observations associated with the largest $\widehat{\mu}_{it}$ fitted values.

It is a fairly common occurrence that the fitted values $\widehat{\mu}_{it}$ will contain tied values. In such cases the number of members in the decile risk groups will not be equal. The grouping is such that members of a given group have similar predicted risk.

We define indicator variables for the first 9 groups and reestimate the model

$$\text{Logit}(p_{it}) = \mathbf{X}\boldsymbol{\beta} + \gamma_1 I_{1it} + \gamma_2 I_{2it} + \cdots + \gamma_9 I_{9it} \qquad (4.55)$$

where I_{kit} is a binary variable indicating whether observation it belongs to group k.

In general, if the original model holds, then $\gamma_1 = \gamma_2 = \cdots \gamma_9 = 0$. Section 4.5 describes the methods for calculating score and Wald tests of coefficients, and the authors (Horton et al.) suggest the score test over the Wald test.

Clearly the Wald test is easier to calculate with standard software, though it does require fitting the alternate model in equation 4.55. With large datasets there will be only small differences in the two tests. See section 4.5 regarding details on when alternate approaches are warranted.

While we have presented the test in its common form, one can use any number of groups G, where the test statistic is distributed as χ^2 with $(G-1)$ degrees of freedom.

Note that it is not always clear when to calculate this test. The model for the Progabide data are a good example of an analysis resulting in tied values. There are only 4 unique fitted values for the 295 observations. It would seem reasonable to define the groups based on these four values, but those indicators are then collinear with the existing covariates already in the model. The analysis of categorical data, when there are only a limited number of covariate patterns, commonly results in a lack of uniquely fitted values.

Returning to the generated quadratic data defined in equation 4.31, we calculate the goodness of fit test using Wald's approach for the linear model

$$\mathbf{y} = \beta_0 + \mathbf{x1}\beta_1 \tag{4.56}$$

Not surprisingly, the results indicate strong evidence that the model does not fit the data, $\chi^2 = 293.27$. Testing the better model

$$\mathbf{y} = \beta_0 + \mathbf{x1}\beta_1 + \mathbf{x12}\beta_2 \tag{4.57}$$

results in $\chi^2 = 5.78$, for which the p-value is 0.7615—indicating no evidence that the model does not fit.

In a study to assess the effect of smoke and pollution on the respiratory symptoms of children, responses were measured, for each child, once a year for 4 years. Covariates included whether a family member smoked, the city in which the child resided, and the age of the child. An exchangeable correlation binomial logit PA-GEE model resulted in

```
GEE population-averaged model              Number of obs      =        100
Group variable:                     id     Number of groups   =         25
Link:                            logit     Obs per group: min =          4
Family:                       binomial                    avg =        4.0
Correlation:              exchangeable                    max =          4
                                           Wald chi2(3)       =       8.26
Scale parameter:                     1     Prob > chi2        =     0.0409
------------------------------------------------------------------------------
     symptom |      Coef.   Std. Err.       z    P>|z|     [95% Conf. Interval]
-------------+----------------------------------------------------------------
        city |  -.0424028   .4908312    -0.09   0.931    -1.004414    .9196087
         age |  -.3200042   .1836975    -1.74   0.082     -.6800447    .0400363
       smoke |   .6519219   .3089066     2.11   0.035      .0464761   1.257368
       _cons |   2.301695   1.780744     1.29   0.196    -1.188499   5.791889
------------------------------------------------------------------------------
```

We evaluate the goodness of fit of the model using 4 groups and obtain the results

```
         chi2(  3) =     1.02
       Prob > chi2 =   0.7962
```

indicating no evidence that there is a lack of fit. The model indicates that symptoms tend to decrease with age and that familial smoking tends to increase the likelihood of respiratory symptoms as the child is exposed.

We emphasize that this test is useful in determining the functional form and specification of covariates for a model. The test is not useful when comparing hypothesized correlation structures. In most cases, the predicted values are rank equal (or nearly so) for different correlation structures in models that are otherwise equivalently specified.

4.5 Testing coefficients in the PA-GEE model

The three standard approaches to constructing test statistics for hypothesis tests are the likelihood ratio test, the Wald test, and the score test. Each of these tests is formed from a quadratic expression of an estimated coefficient vector, and an estimate of the variance of the coefficients. Introductory modeling texts typically focus on likelihood-based models for which a discussion of test construction focuses on only one approach. Here we wish to differentiate 3 possible approaches to building each of the tests.

Throughout, we assume that $(p \times 1)$ coefficient vector β^{T} may be written as the augmented vector $(\gamma^{\mathrm{T}}, \delta^{\mathrm{T}})$ where the $(r \times 1)$ vector γ^{T} is the first r component of β^{T}, and the $((p-r) \times 1)$ vector δ^{T} are the remaining components. Further, we let $\mathbf{V}_{\mathrm{MS}}$ denote the modified sandwich estimate of variance and $\mathbf{V}$ denote the naive variance. The hypothesis test of interest is

$$H_0: \quad \gamma = \gamma_0 \qquad (4.58)$$

where γ_0 is the hypothesized value of γ.

In presenting the derivations and formulas for testing PA-GEE models, we let $\widehat{\beta}$ denote the estimated coefficient vector, $\widehat{\beta}_{\mathrm{I}}$ denote the estimated coefficient vector for the independence model, and $\widehat{\beta}_{\mathrm{IC}}$ denote the estimated coefficient vector for the independence model subject to the constraints of the hypothesis test.

Approach 1 is analogous to a likelihood-based modeling approach. The test statistic is constructed using the naive variance estimate and assumed to follow a χ^2 distribution with r degrees of freedom. Approach 2 is characterized by constructing the test statistic using a robust variance estimate. The test statistic is also assumed to follow a χ^2 distribution with r unadjusted degrees of freedom (*generalized hypothesis test*). Approach 3 constructs the test statistic using the naive variance estimate. The test statistic is assumed to follow the χ^2 distribution with r^* adjusted degrees of freedom (*working hypothesis test*).

Most software packages support the first two approaches to the construction of hypothesis tests. The third approach is not currently supported. In all of these approaches to the construction of a test of the hypothesis, the PA-GEE model is estimated using the user-specified correlation structure. If a test is constructed using results from the independence model, it is called a *naive hypothesis test*.

Rotnitzky and Jewell (1990) present derivations and extensions of the working and naive hypothesis tests that are discussed here for the PA-GEE model. In the following subsections we present the derivations of the hypothesis tests and use various examples to illustrate the tests.

4.5.1 Likelihood ratio tests

The usual likelihood ratio test can not be applied to PA-GEE models since there is no associated likelihood underlying the model. However, a naive likelihood ratio test may be calculated under the associated independence model. The naive likelihood ratio test is then a comparison of the likelihood $\mathcal{L}(\widehat{\beta}_\mathrm{I})$ for fitting the unconstrained independence model to the likelihood $\mathcal{L}(\widehat{\beta}_\mathrm{IC})$ for fitting the constrained independence model.

$$T^*_\mathrm{LR} = 2\Big[\mathcal{L}(\widehat{\beta}_\mathrm{I}) - \mathcal{L}(\widehat{\beta}_\mathrm{IC})\Big] \quad (4.59)$$

In the usual likelihood setting, the test statistic is distributed χ^2 with degrees of freedom equal to r. In the case of PA-GEE models, the test statistic is still distributed χ^2, but with an adjusted degrees of freedom parameter calculated from the data. The test statistic may be written

$$T^*_\mathrm{LR} = \sum_{j=1}^{r} d_j \chi_1^2 \quad (4.60)$$

where $d_1 \geq d_2 \geq \cdots \geq d_r$ are the (ordered) eigenvalues of $P \approx P_0^{-1} P_1$

$$P_0 = \frac{1}{n}\sum_{i=1}^{n} \widetilde{X}_i \Delta_i A_i \Delta_i X_i \quad (4.61)$$

$$P_1 = \frac{1}{n}\sum_{i=1}^{n} \widetilde{X}_i \Delta_i \mathrm{V}(Y_i) \Delta_i \widetilde{X}_i \quad (4.62)$$

where for $r < p$,

$$\widetilde{X}_i = X_i^{(1)} - X_i^{(2)} \left(\sum_{i=1}^{n} X_i^{(2)\mathrm{T}} \Delta_i A_i \Delta_i X_i^{(2)}\right)^{-1} \left(\sum_{i=1}^{n} X_i^{(2)\mathrm{T}} \Delta_i A_i \Delta_i X_i^{(1)}\right) \quad (4.63)$$

and $\widetilde{X}_i = X_i$ for $r = p$.

TESTING COEFFICIENTS IN THE PA-GEE MODEL

When we are testing a single covariate, the value of d_1 simplifies to the ratio of the variance of the tested coefficient in the PA-GEE model to the variance of the tested coefficient in the independence model. In the calculation of this ratio, the modified sandwich estimates of variance for each model should be used.

Using the Progabide data, we can test the hypothesis that the coefficient of timeXprog interaction is zero using the simplification described.

Fitting the exchangeable correlation Poisson model to the data without constraints results in

```
GEE population-averaged model                   Number of obs      =        295
Group variable:                          id     Number of groups   =         59
Link:                                   log     Obs per group: min =          5
Family:                             Poisson                    avg =        5.0
Correlation:                   exchangeable                    max =          5
                                                Wald chi2(3)       =       0.92
Scale parameter:                          1     Prob > chi2        =     0.8203

                       (standard errors adjusted for clustering on id)
------------------------------------------------------------------------------
             |             Semi-robust
    seizures |      Coef.   Std. Err.      z    P>|z|     [95% Conf. Interval]
-------------+----------------------------------------------------------------
        time |   .111836    .1169256     0.96   0.339    -.1173339    .3410059
    progabide|   .0275345   .2236916     0.12   0.902    -.410893     .465962
    timeXprog|  -.1047258   .2152769    -0.49   0.627    -.5266608    .3172092
       _cons |   1.347609   .1587079     8.49   0.000     1.036547    1.658671
    lnPeriod |   (offset)
------------------------------------------------------------------------------
```

The modified sandwich variance of the timeXprog coefficient is .0463.

Fitting an independence model to the data results in

```
Generalized linear models                       No. of obs         =        295
Optimization     : ML: Newton-Raphson           Residual df        =        291
                                                Scale param        =          1
Deviance         =  3575.423602                 (1/df) Deviance    =   12.28668
Pearson          =  5726.792994                 (1/df) Pearson     =    19.6797

Variance function: V(u) = u                     [Poisson]
Link function    : g(u) = ln(u)                 [Log]
Standard errors  : Sandwich

Log likelihood   = -2318.503321                 AIC                =   15.74579
BIC              =     3552.6757

------------------------------------------------------------------------------
             |              Robust
    seizures |      Coef.   Std. Err.      z    P>|z|     [95% Conf. Interval]
-------------+----------------------------------------------------------------
        time |   .111836    .1943749     0.58   0.565    -.2691317    .4928038
    progabide|   .0275345   .2221647     0.12   0.901    -.4079003    .4629693
    timeXprog|  -.1047258   .2946478    -0.36   0.722    -.6822248    .4727732
       _cons |   1.347609   .1576245     8.55   0.000     1.038671    1.656548
    lnPeriod |   (offset)
------------------------------------------------------------------------------
```

The modified sandwich variance for the `progabide` coefficient is .0868 for the independence model. Log-likelihood values for independence models with (shown above) and without (not shown) the `progabide` covariate are

$$\mathcal{L}_{\text{full}} = -2318.503321 \tag{4.64}$$
$$\mathcal{L}_{\text{subset}} = -2319.800533 \tag{4.65}$$

The naive likelihood ratio test statistic is therefore equal to 2.59 and the adjusted degrees of freedom is $.0463/.0868 = .5534$. The results of the naive likelihood ratio test are thus summarized as

$$T^*_{\text{LR}} = 2.59 \tag{4.66}$$
$$p = .0479 \tag{4.67}$$

comparing the test statistic to a χ^2 random variable with .5534 degrees of freedom. The test rejects the hypothesis that the coefficient on `timeXprog` is 0 at a $\alpha = .05$ level of significance.

4.5.2 Wald tests

Most software packages will allow Wald-type testing of coefficients after model estimation using either the naive or the modified sandwich estimate of variance. Test statistics are typically calculated using Wald tests without adjusting the degrees of freedom. When it is said that the degrees of freedom are unadjusted, it means that there is no *a priori* algorithm that calculates an adjustment from data. However, some software packages will make a different kind of adjustment when the modified sandwich estimate of variance is singular (when there are more covariates than panels). Using the modified sandwich estimate of variance results in the generalized Wald test statistic

$$T_W = n(\widehat{\gamma} - \gamma_0)^T V_{\text{MS}}^{-1}(\widehat{\gamma} - \gamma_0) \tag{4.68}$$

In most cases software packages that allow post-estimation Wald tests will use degrees of freedom equal to r. As always, when using the modified sandwich estimate of variance to construct test statistics, we should ensure that there are less covariates than panels; otherwise the modified sandwich estimate of variance is singular.

The generalized Wald test can easily be performed as a post-estimation command using the Progabide data. First, the output of fitting a PA-GEE model is obtained by

```
GEE population-averaged model          Number of obs      =        295
Group variable:                   id   Number of groups   =         59
Link:                            log   Obs per group: min =          5
Family:                      Poisson                  avg =        5.0
Correlation:            exchangeable                  max =          5
                                       Wald chi2(3)       =       0.92
Scale parameter:                   1   Prob > chi2        =     0.8203
```

```
                              (standard errors adjusted for clustering on id)
                                Semi-robust
    seizures |    Coef.      Std. Err.       z      P>|z|      [95% Conf. Interval]
-------------+----------------------------------------------------------------------
        time |   .111836     .1169256      0.96    0.339      -.1173339    .3410059
    progabide |   .0275345    .2236916      0.12    0.902       -.410893    .465962
   timeXprog |  -.1047258    .2152769     -0.49    0.627      -.5266608    .3172092
       _cons |  1.347609     .1587079      8.49    0.000       1.036547    1.658671
    lnPeriod |  (offset)
```

Test results evaluating whether progXtime is equal to 0 provides statistics of

$$T_W = 0.24 \qquad (4.69)$$
$$p = .6266 \qquad (4.70)$$

comparing the test statistic to a χ^2 random variable with 1 degree of freedom. The test fails to reject the hypothesis that the coefficient on timeXprog is 0 at the $\alpha = .05$ level. It is not uncommon to reach different conclusions using different types of χ^2 tests.

In this particular case, the generalized Wald test is actually already part of the output listed for the model. In the model output, the test that each column is 0 is presented as a test with a normally distributed test statistic. The χ^2 statistic is equal to the square of this statistic ($.24 = (-.49)^2$) and the p-values are the same.

An alternative to the generalized Wald test is the working Wald test. In this approach, we construct a test that uses the naive variance estimate to avoid the singularity problems that might arise using the modified sandwich estimate of variance.

The working Wald test statistic is defined as

$$T_W^* = n(\widehat{\gamma} - \gamma_0)^T V^{-1}(\widehat{\gamma} - \gamma_0) \qquad (4.71)$$

This approach assumes that the correlation parameters, α, describe the true structure of the panels. Regardless of whether this is in fact true, it is still possible to describe the degrees of freedom of the test statistic.

The working test may be written

$$T_W^* = \sum_{j=1}^{r} c_j \chi_1^2 \qquad (4.72)$$

where $c_1 \geq c_2 \geq \cdots \geq c_r$ are the (ordered) Eigenvalues of $Q \approx Q_0^{-1} Q_1$

$$Q_0 = \frac{1}{n} \sum_{i=1}^{n} \widetilde{D}_i V_i^{-1} D_i \qquad (4.73)$$

$$Q_1 = \frac{1}{n} \sum_{i=1}^{n} \widetilde{D}_i V_i^{-1} \text{Cov}(Y_i) V_i^{-1} \widetilde{D}_i \qquad (4.74)$$

where for $r < p$,

$$\widetilde{D}_i = D_i^{(1)} - D_i^{(2)} \left(\sum_{i=1}^n D_i^{(2)\mathrm{T}} V_i^{-1} \widetilde{D}_i^{(2)} \right)^{-1} \left(\sum_{i=1}^n D_i^{(2)\mathrm{T}} V_i^{-1} \widetilde{D}_i^{(1)} \right) \quad (4.75)$$

and $\widetilde{D}_i = D_i$ for $r = p$.

4.5.3 Score tests

The development of score hypothesis tests follows the development seen in the previous section. The generalized score test substitutes the modified sandwich estimate of variance for the naive variance to obtain the test statistic

$$T_\mathrm{S} = \frac{1}{n} \Psi_\gamma(\gamma_0, \widehat{\delta}(\gamma_0))^\mathrm{T} V_{\mathrm{MS},\gamma} \Psi_\gamma(\gamma_0, \widehat{\delta}(\gamma_0)) \quad (4.76)$$

where $\Psi_\gamma()$ is the constrained estimating equation of the PA-GEE model of interest.

In the case that the number of covariates exceeds the number of panels, we can consider the working score test as

$$T_\mathrm{S}^* = \frac{1}{n} \Psi_\gamma(\gamma_0, \widehat{\delta}(\gamma_0))^\mathrm{T} V_\gamma \Psi_\gamma(\gamma_0, \widehat{\delta}(\gamma_0)) \quad (4.77)$$

where (as in the case of the working Wald test)

$$T_\mathrm{S}^* = T_\mathrm{W}^* = \sum_{j=1}^r c_j \chi_1^2 \quad (4.78)$$

and where $c_1 \geq c_2 \geq \cdots \geq c_r$ are the (ordered) eigenvalues of $Q \approx Q_0^{-1} Q_1$ (defined in equations 4.73 and 4.74).

4.6 Assessing the MCAR assumption of PA-GEE models

It was mentioned in the previous chapter that the PA-GEE models depend on an assumption that is a special case of MCAR for missing data. Here, we discuss various techniques for assessing the validity of this assumption.

Section 3.7 included an illustration of the various patterns of missing data that might be seen in data. We shall use this illustration for a first look at the patterns.

Fanurik, Zeltzer, Roberts, and Blount (1993) present an example of a study on pain tolerance in children. Each child participating in the study placed his or her hands in cold water for as long as possible. The response variable for the pain tolerance proxy is the log of time in seconds, lntime, that each child was able to keep his or her hands in the water. Each child participated in the trials by repeating the experiment 4 times. The children were classified as either attenders or distractors, where these terms related to the child's coping style cs. The attenders tended to concentrate on either the experimental apparatus or on their hands, while the distractors tended to concentrate on unrelated things (e.g., home or school). Three baseline measurements were collected for each child before a counseling session trt was held. These counseling sessions were randomized to teach the attending coping style, the distractor coping style, or the control (no advice). A fourth measurement was collected after the counseling session. The authors expected that altering a child's natural coping style would impede his or her performance, while the control counseling session was expected to have no effect on the performance. The study included 64 children with 11 missing data.

There would be 256 observations if the study were complete. Due to various reasons, 11 total observations are missing from 6 of the 64 children. Our purpose here is not to propose a full analysis of the data, but rather to investigate the missing data mechanisms related to PA-GEE modeling.

The illustration below provides a graph of the missing data.

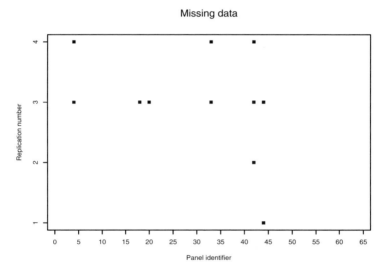

Squares mark missing data for the response variable in a dataset with 64 panels with 4 repeated measures per panel.

To test the PA-GEE MCAR assumption

$$P(\mathbf{R}_i|\mathbf{Y}_i, \mathbf{X}_i, \boldsymbol{\beta}_i) = P(\mathbf{R}_i) \qquad (4.79)$$

a binary variable is created to indicate the groups with missing observations. A t-test is performed on the other covariates. This investigation can be used for each covariate at each replication number. For the pediatric pain data at the third replication, the coping style results are

Two-sample t test with equal variances

```
---------+--------------------------------------------------------------
   Group |    Obs        Mean     Std. Err.    Std. Dev.    [95% Conf. Interval]
---------+--------------------------------------------------------------
       0 |      6         1.5     .2236068     .5477226     .9252004    2.0748
       1 |     58         1.5     .0662266     .5043669     1.367383    1.632617
---------+--------------------------------------------------------------
combined |     64         1.5     .0629941     .5039526     1.374116    1.625884
---------+--------------------------------------------------------------
    diff |                  0     .2178535                  -.4354829   .4354829
---------+--------------------------------------------------------------
Degrees of freedom: 62

                       Ho: mean(0) - mean(1) = diff = 0

    Ha: diff < 0              Ha: diff ~= 0              Ha: diff > 0
      t =   0.0000              t =   0.0000              t =   0.0000
  P < t =   0.5000          P > |t| =   1.0000          P > t =   0.5000
```

The results for counseling session treatment at the third replication are:

Two-sample t test with equal variances

```
---------+--------------------------------------------------------------
   Group |    Obs        Mean     Std. Err.    Std. Dev.    [95% Conf. Interval]
---------+--------------------------------------------------------------
       0 |      6    2.333333     .3333333     .8164966     1.476473    3.190194
       1 |     58    1.948276     .1055691     .8039904     1.736877    2.159674
---------+--------------------------------------------------------------
combined |     64    1.984375     .1008205     .8065641     1.782901    2.185849
---------+--------------------------------------------------------------
    diff |           .3850575     .3452229                  -.3050332   1.075148
---------+--------------------------------------------------------------
Degrees of freedom: 62

                       Ho: mean(0) - mean(1) = diff = 0

    Ha: diff < 0              Ha: diff ~= 0              Ha: diff > 0
      t =   1.1154              t =   1.1154              t =   1.1154
  P < t =   0.8655          P > |t| =   0.2690          P > t =   0.1345
```

In both cases, there is no significant difference. Here, we are interested only in the covariates for coping style and counseling session. There are $p(p-1)$ possible tests that can be performed in a model with p covariates and T replications. Care must be taken regarding accumulated Type I errors.

Little (1988) presents a single test for assessing MCAR. In this approach, a vector of means is computed for the construction of a χ^2 statistic. The statistic is defined as

$$d^2 = \sum_{i=1}^{n} n_i (\overline{\mathbf{y}}_i - \widehat{\boldsymbol{\mu}}_i) \widehat{\boldsymbol{\Sigma}}^{-1} (\overline{\mathbf{y}}_i - \widehat{\boldsymbol{\mu}}_i)^{\mathrm{T}} \qquad (4.80)$$

where $\overline{\mathbf{y}}_i$ is a vector of values for the observed variables in panel i. $\widehat{\boldsymbol{\mu}}$ and $\widehat{\boldsymbol{\Sigma}}$

are the maximum likelihood estimates assuming that the panels are iid normal and the missing data process is ignorable.

The distribution of the test statistic can be complex when related to monotone missing data (dropouts). However, a common type of data situation involves a single follow-up repeated measurement ($n_i = 2$), for which the test statistic can be computed as

$$d^2 = \frac{\text{SSB}}{\text{MST}} = \frac{(n-1)F}{(n-2+F)} \tag{4.81}$$

and where SSB and MST are the between sum of squares and total mean square from an analysis of variance of Y_1 on the missing data pattern. F is the test statistic for the ANOVA model. In this case, there are only two missing data patterns: y_{i1}, y_{i2} are both observed, and y_{i1} is observed while y_{i2} is missing.

Section 3.7 presents a sample dataset of the asthma status of white children. We can apply the above test to these data. An analysis of variance of y_{i1} on the missing data pattern is given by

```
         Number of obs =    2566     R-squared     =  0.0012
         Root MSE      = .210256     Adj R-squared =  0.0008

  Source |  Partial SS    df      MS             F     Prob > F
---------+------------------------------------------------------
   Model |  .133234469     1   .133234469        3.01    0.0827
         |
       g |  .133234469     1   .133234469        3.01    0.0827
         |
Residual |  113.348059  2564   .044207511
---------+------------------------------------------------------
   Total |  113.481294  2565    .04424222
```

The test statistic for the monotone missing data is then

$$d^2 = 3.0088 \tag{4.82}$$

Since the F statistic here is equal to the square of a t statistic, this test (for this particular case of 2 outcomes) is the same as the t-test previously described.

4.7 Summary

Standard exploratory data analysis (EDA) techniques should be used with panel datasets. Plots of the raw data can be constructed, with particular attention to illustrations depicting both the panel nature of the data and the repeated measures identifiers. These types of plots assist the analyst in identifying dependence on time as well as on the panels. In addition, standard GLM-type plots of the Pearson residuals versus the linear predictor and the Pearson residuals versus the variance are used to assess model adequacy.

Model assessment is based on graphical, as well as statistical, points of view. Graphs of influence and leverage uncover outliers in the data that may not be noticed otherwise. These outliers should be investigated for data integrity, and statistical measures should be calculated to provide a measure of the effect of these outliers on the fitted models.

There are two deletion diagnostic approaches to measuring the effects of outliers on fitted models: deleting individual observations and deleting entire panels of observations. We use both techniques since the two deletion diagnostic criteria do not summarize the same information in the data.

Model criterion measures are provided to assess overall model goodness of fit. The QIC measure is a particularly useful tool for choosing the best correlation structure in a PA-GEE model. Similarly, the QIC_u measure is used for model selection.

Standard model criterion measures, such as R^2, are available for panel data models. These ubiquitous measures have a long history in statistical models. They have a clear interpretation for OLS linear regression, and researchers have produced a long list of references for extending the measure to nonlinear models. However, the R^2 measure can be difficult to interpret for nonlinear models, and experience may be an analyst's best ally in terms of interpreting the magnitude of R^2 in a particular situation.

We recommend using the generalized forms of the various tests for the majority of data analysis situations. These are the easiest tests to perform and the tests are supported in many standard software packages. Interpretation is easy, but the analyst must be aware of the fact that the modified sandwich estimate of variance has a rank that depends on the number of panels. If the number of panels is less than the number of covariates in the model, or the number of panels is not too much larger, the working versions of the tests are preferred over the generalized versions of the tests. However, working tests must be programmed by the analyst.

This chapter attempts to illustrate and catalog techniques for assessing model adequacy. We pay particular attention to those criteria for PA-GEE models since those are techniques most notably missing from software documentation and other texts. We refrained from presenting full analyses for the various datasets in order to focus attention on each technique. A complete analysis would use several of the techniques listed.

4.8 Exercises

1. In a study of the efficacy of a new drug treatment for depression, patients receiving the new drug are likely to drop out of the study if they experience dramatic improvement. Patients receiving the placebo are also likely to drop out of the study if there is no observed response. Is the characterization of missing data amenable to modeling by PA-GEE?

2. The QIC measure is used to choose the best correlation structure among PA-GEE models with the same covariates. Show that apart from a normalizing term for the quasilikelihood, the QIC measure is equal to the AIC for an independence PA-GEE model implying a likelihood proper; that is, for an independence PA-GEE model which specifies a variance function from a member of the exponential family of distributions.

3. Deletion diagnostics can be calculated by either deleting individual observations, or by deleting entire panels of observations. Discuss which calculation method you would prefer in assessing a PA-GEE model.

4. Verify the output provided in Table 4.4 and Table 4.6 using the Progabide data in section 5.2.3.

5. An example illustrated the fit of the quasivariance $V(\mu) = \mu^2(1-\mu)^2$ (see section 4.2.3). Derive the associated deviance.

CHAPTER 5

Programs and Datasets

This chapter presents some of the datasets used in this text together with samples of using the mentioned software packages. In addition to providing sample code for fitting the GEE models discussed in this text, we also provide sample code for fitting alternative likelihood-based models where appropriate. This section will not teach the reader how to use a particular software package, but will provide samples from which other analyses can be produced. We pay particular attention to the various options available in each package in order to highlight the relatively minor differences between them.

Samples are presented by means of input files which can be run in "batch" mode. Readers can then interactively enter the commands; however, we suggest first running the programs in their entirety and then reading the results.

The PA-GEE model has the best support across packages. There are some minor differences in the default behavior for the packages that we highlight in the following programs. We include comments to assist the reader in understanding the choices of both the analysis and the selected options.

The datasets listed in this chapter are available in tab-delimited plain text format from:

http://www.crcpress.com/e_products/downloads/download.asp?cat_no=C3073

5.1 Programs

The following pages include programs for fitting various models and calculating various diagnostic statistics. These programs can be run in batch mode and illustrate each of the software packages used in this text.

5.1.1 Fitting PA-GEE models in Stata

```
* This program uses the data listed in section 1.3.4
* The program assumes the data are available as 'qicdata.txt'
infile id t x1 x2 x3 ei xx ys y using qicdata.txt

* Fit an exchangeable correlation logistic regression GEE-PA
* using divisor (n) for the dispersion
xtgee y x1 x2 x3, i(id) t(t) fam(bin) corr(exch)

* Fit an exchangeable correlation logistic regression GEE-PA
* using divisor (n-p) for the dispersion
xtgee y x1 x2 x3, i(id) t(t) fam(bin) corr(exch) nmp

* Fit an exchangeable correlation probit regression GEE-PA
* using divisor (n) for the dispersion
xtgee y x1 x2 x3, i(id) t(t) fam(bin) link(probit) corr(exch)

exit

Notes:
  Variance specification    Variance
  ----------------------    --------
  fam(bin)                  mu(1-mu)
  fam(bin k)                mu(1-mu/k)
  fam(bin variable)         mu(1-mu/[variable_it])
  fam(gaussian)             1
  fam(gamma)                mu^2
  fam(igaussian)            mu^3
  fam(nbinomial k)          mu + k*mu^2
  fam(poisson)              mu

  Link specification
  ------------------
  link(identity)            mu
  link(cloglog)             ln(-ln(1-mu))
  link(log)                 ln(mu)
  link(logit)               ln(mu/(1-mu))
  link(nbinomial)           ln(mu/(mu+1/k))
  link(opower a)            [mu/(1-mu)^a - 1] / a
  link(power a)             mu^a
  link(probit)              InvPhi(mu)
  link(reciprocal)          1/mu

  nmp  --  option to allow estimation of the dispersion parameter
           with denominator (n-p) instead of the default (n)
```

5.1.2 Fitting PA-GEE models in SAS

```
options ls = 80 ;
data qic ;
        infile 'qicdata.txt' ;
        input id t x1 x2 x3 ei xx ys y ;

/* Fit an exchangeable correlation logistic regression GEE-PA
   using divisor (n-p) for the dispersion */
proc genmod data=qic ;
        model y = x1 x2 x3 / dist=bin link=logit ;
        repeated subject=id / corr=exch ;
run ;
quit ;

/*
Notes:
  Variance specification       Variance
  ----------------------       --------
  dist=bin                     mu(1-mu)
  dist=gamma                   mu^2
  dist=igaussian               mu^3
  dist=multinomial             ....
  dist=negbin                  mu + k*mu^2
  dist=normal                  1
  dist=poisson                 mu

  variance specification is an option in the MODEL statement

  Link specification
  ------------------
  link=cumcll               ln(-ln(1-mu1))  , ln(-ln(1-(mu1+mu2)))          , ...
  link=cumlogit             ln(mu1/(1-mu1)), ln((mu1+mu2),/(1-(mu1+mu2))), ...
  link=cumprobit            InvPhi(mu1)    , InvPhi(mu1+mu2)               , ...
  link=cloglog              ln(-ln(1-mu))
  link=identity             mu
  link=log                  ln(mu)
  link=logit                ln(mu/(1-mu))
  link=power(a)             mu^a
  link=probit               InvPhi(mu)

  link specification is an option in the MODEL statement

  Correlation structure
  ---------------------
  corr=XXXXX        GEE-PA
  logor=XXXXX       ALR

  correlation specification is an option in the REPEATED statement

  V6CORR -- option to allow estimation of the dispersion parameter with
            (n) denominator instead of the default (nmp). This option
            is specified in the REPEATED statement.
*/
```

5.1.3 Fitting PA-GEE models in S-PLUS

```
#
# Uses geex package from http://lib.stat.cmu.edu/S/geex
#

# This program uses the data listed in section 1.3.4
# The program assumes the data are available as 'qicdata.txt'
qicdata <- scan("qicdata.txt",list(id=0,t=0,x1=0,x2=0,
                x3=0,ei=0,xx=0,ys=0,y=0))

# Fit an exchangeable correlation logistic regression GEE-PA
# using divisor (n) for the dispersion
gee.exc <- gee(formula=y~x1+x2+x3,
               family=binomial(link=logit),
               correlation="compoundsymmetric",
               data=qicdata,
               subject=id,
               repeated=t)


# Fit an exchangeable correlation probit regression GEE-PA
# using divisor (n) for the dispersion
gee.exc <- gee(formula=y~x1+x2+x3,
               family=binomial(link=probit),
               correlation="compoundsymmetric",
               data=qicdata,
               subject=id,
               repeated=t)


#
# Notes:
#    Variance specification           Variance
#    ----------------------           --------
#    family=binomial(link=lnkname)    mu(1-mu)
#    family=gaussian(link=lnkname)       1
#    family=gamma(link=lnkname)         mu^2
#    family=igaussian(link=lnkname)     mu^3
#    family=poisson(link=lnkname)       mu
#
#
#    Correlation specification
#    -------------------------
#    independent
#    unstructured
#    compoundsymmetric
#    autoregressiveI
#    dependentI
#
```

5.1.4 Fitting ALR models in SAS

```
options ls = 80 ;
data qic ;
        infile 'qicdata.txt' ;
        input id t x1 x2 x3 ei xx ys y ;

/* Fit an exchangeable correlation logistic regression GEE-PA
   using divisor (n-p) for the dispersion */
proc genmod data=qic ;
        model y = x1 x2 x3 / dist = bin ;
        repeated subject=id / logor=exch ;
run ;
quit ;

/*
Notes:
  Variance specification      Variance
  ----------------------      --------
  dist=bin                    mu(1-mu)
  dist=gamma                  mu^2
  dist=igaussian              mu^3
  dist=multinomial            ....
  dist=negbin                 mu + k*mu^2
  dist=normal                 1
  dist=poisson                mu

  Link specification
  ------------------
  link=cumcll                 ln(-ln(1-mu1)) , ln(-ln(1-(mu1+mu2)))       , ...
  link=cumlogit               ln(mu1/(1-mu1)), ln((mu1+mu2),/(1-(mu1+mu2))), ...
  link=cumprobit              InvPhi(mu1)    , InvPhi(mu1+mu2)            , ...
  link=cloglog                ln(-ln(1-mu))
  link=identity               mu
  link=log                    ln(mu)
  link=logit                  ln(mu/(1-mu))
  link=power(a)               mu^a
  link=probit                 InvPhi(mu)

  Correlation structure
  ---------------------
  corr=XXXXX           GEE-PA
  logor=XXXXX          ALR

  V6CORR -- option to allow estimation of the dispersion parameter with
           (n) denominator instead of the default (nmp).  This option
           is specified in the REPEATED statement.
*/
```

5.1.5 Fitting PA-GEE models in SUDAAN

```
/* This code is for use where SUDAAN is a callable PROC
          from SAS */

options ls = 80 ;
data mlog ;
        infile 'mlog.txt' ;
        input id t y x1 x2 ;

/* Fit an exchangeable correlation generalized
          logistic regression GEE-PA */

proc multilog data=mlog R=exchangeable ;
       nest _one_ id ;
       weight _one_ ;
       subgroup y ;
       levels 3 ;
       model y = x1 x2 / genlogit ;
run ;

/* Fit an exchangeable correlation cumulative
          logistic regression GEE-PA */

proc multilog data=mlog R=exchangeable ;
       nest _one_ id ;
       weight _one_ ;
       subgroup y ;
       levels 3 ;
       model y = x1 x2 / cumlogit ;
run ;
quit ;

/*
Notes:
  R=independent          independence model
  R=exchangeable         GEE-PA exchangeable correlation

  There are only the above two correlation structures.

*/
```

PROGRAMS

5.1.6 Calculating QIC in Stata

```
* This program uses the data listed in section 1.3.4
* The program assumes the data are available as 'qicdata.txt'
infile id t x1 x2 x3 ei xx ys y using qicdata.txt

* First, we need the naive variance matrix from
* the independence model.
xtgee y x1 x2 x3, i(id) t(t) fam(bin) corr(ind)

* We save the variance matrix, and calculate the inverse.
matrix A  = e(V)
matrix Ai = syminv(A)

* We write a program that runs the same model with a
* specified correlation structure. We need the
* robust variance matrix from this model to
* calculate the QIC criterion measure.
capture program drop qicm
program define qicm
                    * Take as an argument the hypothesized correlation structure
          args corr
          quietly {
                    capture quietly xtgee y x1 x2 x3, i(id) t(t) /*
                        */ fam(bin) corr('corr') robust
                    if (_rc != 0) {
                           * Exit if the model fails to converge
                            exit
                    }
                           * Calculate trace(Ai*V)
                    matrix V   = e(V)
                    matrix T   = Ai*V
                    matrix t   = trace(T)
                    scalar off = t[1,1]
                           * Calculate the remaining term for this
                           * particular model (it is model dependent)
                    tempvar ql
                    quietly predict double mu, mu
                    quietly generate double 'ql' = (y*log(mu/(1-mu)) + log(1-mu))
                    quietly summarize 'ql', meanonly
          }
          display in green "QIC = " in yellow 2*(off-r(sum))
end

* Calculate the QIC statistic for a variety of alternate
* correlation structures. There should be only one argument,
* so if the correlation structure has an optional argument,
* we must enclose the specification in quotes.

qicm ind
qicm exch
qicm "ar 1"
qicm "ar 2"
qicm unst
qicm "sta 2"
```

5.1.7 Calculating QICu in Stata

```
*
* This do-file assumes that the Progabide data are available
* as a Stata dataset named progdata.dta.
*
use progdata, clear

capture program drop qicu
program define qicu
        quietly {
                local p = e(df_m) + 1
                tempvar xb ql
                predict double `xb', xb
                generate double `ql' = seizures*`xb' - exp(`xb') - /*
                        */ seizures*(log(seizures)-1)
                summarize `ql'
                local qicu = -2*r(sum) + 2*`p'
        }
        display in green "QIC_u = " in yellow %8.4f `qicu'
end

* Store the common options in a macro to save typing

local opts "offset(lnPeriod) fam(poiss) corr(exch) i(id) nolog"

* Run each model and calculate QICu for the model.  This will
* produce a lot of output including the results of fitting
* each model along with the QICu criterion measure.

xtgee seizures time, `opts'
qicu

xtgee seizures progabide, `opts'
qicu

xtgee seizures timeXprog, `opts'
qicu

xtgee seizures time progabide, `opts'
qicu

xtgee seizures time timeXprog, `opts'
qicu

xtgee seizures progabide timeXprog, `opts'
qicu

xtgee seizures time progabide timeXprog, `opts'
qicu

xtgee seizures , `opts'
qicu
```

PROGRAMS

5.1.8 Graphing the residual runs test in S-PLUS

```
residrun <- function(ehat, yhat, subtitle="") {

# The graph will reflect the current sort order of the
# residuals and fitted values.  Calling program is responsible
# for ordering the data.

        sgn <- sign(ehat)
        num <- 1:length(ehat)

        plot(xsg, sgn,
                type="s",
                main="Graphical Illustration of Residual Runs",
                sub=subtitle,
                xlab="Fitted values",
                ylab="Sign(Pearson residuals)",
                xlim=c(0,length(ehat)),
                ylim=c(-1,1)
        )
}
```

5.1.9 Using the fixed correlation structure in Stata

```
capture program drop calcR
program define calcR
        capture drop res mm
        predict double res
        replace res   = y-res
        generate double mm = res*res
        summarize mm, meanonly
        scalar phi    = r(mean)
        scalar alpha  = 0
        local i 1
        local n 0
        while 'i' <= $G {
                local ioff = ('i'-1) * $Ni
                local j 1
                while 'j' < $Ni {
                        local joff = 'ioff' + 'j'
                        local koff = 'joff' + 1
                        scalar rij = res['joff']
                        scalar rik = res['koff']
                        scalar alpha = alpha + rij*rik
                        local n = 'n'+1
                        local j = 'j'+2
                }
                local i = 'i'+1
        }
        scalar rr = (alpha/'n')/phi
        matrix R = ( 1,rr, 0, 0, 0, 0, 0, 0\       /*
                */ rr, 1, 0, 0, 0, 0, 0, 0\        /*
                */ 0, 0, 1,rr, 0, 0, 0, 0\         /*
                */ 0, 0,rr, 1, 0, 0, 0, 0\         /*
                */ 0, 0, 0, 0, 1,rr, 0, 0\         /*
                */ 0, 0, 0, 0,rr, 1, 0, 0\         /*
                */ 0, 0, 0, 0, 0, 0, 1,rr\         /*
                */ 0, 0, 0, 0, 0, 0,rr, 1)
end
capture program drop Iter
program define Iter
        global Ni = 8
        global G  = 10
        xtgee y x1 x2, i(id) t(t) corr(ind) nolog
        local i 1
        while 'i' <= 10 {
                matrix R = e(R)
                matrix b = e(b)
                capture xtgee y x1 x2, i(id) t(t) corr(fixed R) iter(2) from(b)
                matrix R = e(R)
                calcR
                local i = 'i'+1
        }
end
* Use the Iter program to iterate several times starting from an
* independence model.  For each iteration, use the currently calculated
* correlation matrix from the calcR program.  Iterate 10 times.
use optdata, clear
quietly Iter
xtgee
matrix list e(R), format(%4.2f)
```

5.1.10 Fitting quasivariance PA-GEE models in S-PLUS

```
#
# Uses geex package from http://lib.stat.cmu.edu/S/geex
#
# The goal is to use the make.family() function to create
# another family (fully supported variance function) for use
# with either the gee() or glm() commands.
#
# First, define the necessary label, variance, and
# deviance functions.  Symbolic math programs can be useful
# for determining the deviance.

binsq.var <- list(
        names="Quasi: mu^2(1-mu)^2",
        variance=substitute(function(mu) {
                mu*mu*(1-mu)*(1-mu)
        }),
        deviance=substitute(function(mu,y,w,residuals=F) {
                dev <- mu
                d0 <- w*mu*(2*mu-1)+w*log(1-mu)
                d1 <- w*(mu-1)*(2*mu-1)+w*log(mu)
                dev <-
                        (
                           (2*mu-1)*(mu-y) + (mu-1)*mu*(2*y-1)*(
                                log(mu/y) - log((mu-1)/(y-1)))
                        ) / (mu*(mu-1))
                dev[y==0] <- d0[y==0]
                dev[y==1] <- d1[y==1]

                if (residuals) {
                        return(y-mu)
                }
                else {
                        return(-2*sum(dev))
                }
        }))

# Make the new quasivariance family

binsq.f <- make.family("Quasi", binomial$link, binsq.var)

# Sample call:
#        Assumes: leafdata is an object holding the variables.
#                 This can be formed using
#
#                 leafdata <- scan("leaf.txt", list(leaf=0,variety=0,
#                         site1=0,site2=0,site3=0,site4=0,site5=0,
#                         site6=0,site7=0,site8=0,site9=0,t=0))
#
#                 if the data are in the 'leaf.txt' file
#
# In the call: specify the created quasivariance function we created

res <- gee(formula=leaf~site1+site2+site3+site4+site5+site6+site7+site8,
           family=binsq.f,
           correlation="compoundsymmetric",
           control=gee.control(tolerance=1e-25,epsilon=1e-14,trace=T),
           data=leafdata,
           subject=variety,
           repeated=t)
```

5.2 Datasets

The following subsections list the data that are used in the examples of the text along with an explanation of the data.

5.2.1 Wheeze data

These data study the effect on health effects of air pollution. These data are a subset of the data in Ware, Docker III, Speizer, and Ferris Jr. (1984). The data include the case number, case; a within subject observation identifier, t; a binary indicator of whether the subject wheezes, wheeze; a binary indicator of whether the observation is in Kingston, kingston; the age of the child in years, age; and a binary indicator of whether the child's mother smokes, smoke. The data are for 16 children where observations were collected for four years at ages 9-12 of the children. The subjects reside in either of the cities Kingston or Portage.

Indicator variables can be constructed for the case. In the text, we include output where _Icase_1 is an indicator that the observation is characterized by case = 1; _Icase_2 indicates that the observation is characterized by case = 2. Indicator variables for the other case identifiers are constructed similarly.

A description of the panel structure of the data is:

```
     case:  1, 2, ..., 16                                  n =        16
        t:  1, 2, ..., 4                                   T =         4
            Delta(t) = 1; (4-1)+1 = 4
            (case*t uniquely identifies each observation)

Distribution of T_i:    min      5%     25%     50%     75%     95%     max
                          4       4       4       4       4       4       4

     Freq.  Percent    Cum. |  Pattern
     ---------------------------+---------
        16   100.00   100.00 |  1111
     ---------------------------+---------
        16   100.00          |  XXXX
```

A summary of the variables is:

```
Variable         |      Mean   Std. Dev.       Min        Max |    Observations
-----------------+--------------------------------------------+----------------
wheeze   overall |   .296875   .4604927         0          1  |    N =      64
         between |             .2918154         0         .75 |    n =      16
         within  |             .3618734   -.453125   1.046875 |    T =       4
                 |                                            |
kingston overall |        .5   .5039526         0          1  |    N =      64
         between |             .5163978         0          1  |    n =      16
         within  |                    0         .5         .5 |    T =       4
                 |                                            |
age      overall |      10.5   1.126872         9         12  |    N =      64
```

DATASETS

```
         between |                  0        10.5     10.5 |   n =     16
         within  |           1.126872        9        12   |   T =      4
                 |                                         |
smoke    overall |  .796875   .6941865        0         2  |   N =     64
         between |            .5493841        0      1.75  |   n =     16
         within  |            .4409586  .046875  1.546875  |   T =      4
```

The data are:

case	t	wheeze	kingston	age	smoke
1	1	1	0	9	0
1	2	1	0	10	0
1	3	1	0	11	0
1	4	0	0	12	0
2	1	1	1	9	1
2	2	1	1	10	2
2	3	0	1	11	2
2	4	0	1	12	2
3	1	1	1	9	0
3	2	0	1	10	0
3	3	0	1	11	1
3	4	0	1	12	1
4	1	0	0	9	0
4	2	1	0	10	0
4	3	1	0	11	0
4	4	0	0	12	1
5	1	0	1	9	0
5	2	0	1	10	1
5	3	0	1	11	1
5	4	0	1	12	1
6	1	0	0	9	0
6	2	0	0	10	1
6	3	0	0	11	1
6	4	0	0	12	1
7	1	0	1	9	1
7	2	0	1	10	1
7	3	0	1	11	0
7	4	0	1	12	0
8	1	0	0	9	1
8	2	0	0	10	1
8	3	0	0	11	1
8	4	0	0	12	2
9	1	1	0	9	2
9	2	0	0	10	2
9	3	0	0	11	1
9	4	0	0	12	1
10	1	0	1	9	0
10	2	0	1	10	0
10	3	0	1	11	0
10	4	0	1	12	1
11	1	1	1	9	1
11	2	0	1	10	0
11	3	1	1	11	0
11	4	1	1	12	0
12	1	0	0	9	1
12	2	0	0	10	0
12	3	0	0	11	0
12	4	0	0	12	0
13	1	0	1	9	1
13	2	1	1	10	0

13	3	1	1	11	1
13	4	1	1	12	1
14	1	0	0	9	1
14	2	0	0	10	2
14	3	0	0	11	1
14	4	1	0	12	2
15	1	0	1	9	1
15	2	0	1	10	1
15	3	0	1	11	1
15	4	1	1	12	2
16	1	1	0	9	1
16	2	1	0	10	1
16	3	0	0	11	2
16	4	0	0	12	1

5.2.2 Ship accident data

This dataset studies the number of accidents reported for ships. Included in the data are a ship identifier id; the number of incidents reported incident; whether the ship operated between 1975 and 1979 op_75_79; whether the ship was in construction between 1965 and 1969 co_65_69; whether the ship was in construction between 1970 and 1974 co_70_74; whether the ship was in construction between 1975 and 1979 co_75_79; the number of months in service during the data collection mon; and the exposure of the ship (natural log of months) exposure. We have included the exposure variable in the data listing, but you can generate this variable as ln(mon) where mon is the number of months that the ship was operating.

Indicator variables can be constructed for the ship identifier. In the text, we include output where _Iship_1 is an indicator that the observation is characterized by ship = 1; _Iship_2 indicates that the observation is characterized by ship = 2. Indicator variables for the other ship types are constructed similarly.

A description of the panel structure of the data is:

```
     ship:  1, 2, ..., 5                              n =      5
        t:  1, 2, ..., 8                              T =      8
            Delta(t) = 1; (8-1)+1 = 8
            (ship*t uniquely identifies each observation)

Distribution of T_i:    min      5%     25%     50%     75%     95%     max
                          8       8       8       8       8       8       8

     Freq.  Percent    Cum. |  Pattern
  -----------------------------+----------
         5   100.00  100.00 |  11111111
  -----------------------------+----------
         5   100.00         |  XXXXXXXX
```

A summary of the variables is:

DATASETS

```
Variable         |     Mean    Std. Dev.      Min        Max  |  Observations
-----------------+--------------------------------------------+----------------
incident overall |      8.9    14.96115         0         58  |  N =       40
         between |             12.7908        1.5     31.625  |  n =        5
         within  |             9.465524    -22.725    35.275  |  T =        8
op_75_79 overall |       .5    .5063697         0          1  |  N =       40
         between |                    0         .5        .5  |  n =        5
         within  |             .5063697         0          1  |  T =        8
                 |                                            |
co_65_69 overall |      .25    .438529          0          1  |  N =       40
         between |                    0        .25        .25 |  n =        5
         within  |             .438529          0          1  |  T =        8
                 |                                            |
co_70_74 overall |      .25    .438529          0          1  |  N =       40
         between |                    0        .25        .25 |  n =        5
         within  |             .438529          0          1  |  T =        8
                 |                                            |
co_75_79 overall |      .25    .438529          0          1  |  N =       40
         between |                    0        .25        .25 |  n =        5
         within  |             .438529          0          1  |  T =        8
                 |                                            |
exposure overall | 7.049255   1.721094    3.806662   10.71179 |  N =       34
         between |            1.516132    5.953964   9.694236 |  n =        5
         within  |            1.014858    4.591827   8.721374 |  T-bar =  6.8
```

where exposure is missing when the ship has not been in service. The data are:

```
ship   t  incident  op_75_79  co_65_69  co_70_74  co_75_79    exposure      mon
  1    1      0         0         1         0         0       4.8441871     127
  1    2      0         1         1         0         0       4.1431347      63
  1    3      3         0         0         1         0       6.9985096    1095
  1    4      4         1         0         1         0       6.9985096    1095
  1    5      6         0         0         0         1       7.3211886    1512
  1    6     18         1         0         0         1       8.1176107    3353
  1    7      0         0         0         0         0           .           0
  1    8     11         1         0         0         0       7.7160153    2244
  2    1     39         0         1         0         0      10.711792    44882
  2    2     29         1         1         0         0       9.7512683   17176
  2    3     58         0         0         1         0      10.261477    28609
  2    4     53         1         0         1         0       9.9218185   20370
  2    5     12         0         0         0         1       8.8627667    7064
  2    6     44         1         0         0         1       9.4802912   13069
  2    7      0         0         0         0         0           .           0
  2    8     18         1         0         0         0       8.8702416    7117
  3    1      1         0         1         0         0       7.0724219    1179
  3    2      1         1         1         0         0       6.313548      552
  3    3      0         0         0         1         0       6.6605751     781
  3    4      1         1         0         1         0       6.5161931     676
  3    5      6         0         0         0         1       6.6631327     783
  3    6      2         1         0         0         1       7.5745585    1948
  3    7      0         0         0         0         0           .           0
  3    8      1         1         0         0         0       5.6131281     274
  4    1      0         0         1         0         0       5.5254529     251
  4    2      0         1         1         0         0       4.6539604     105
  4    3      0         0         0         1         0       5.6629605     288
  4    4      0         1         0         1         0       5.2574954     192
  4    5      2         0         0         0         1       5.8550719     349
  4    6     11         1         0         0         1       7.0967214    1208
```

```
          4    7       0       0       0       0       0         .            0
          4    8       4       1       0       0       0    7.6260828      2051
          5    1       0       0       1       0       0    3.8066625        45
          5    2       0       1       1       0       0         .            0
          5    3       7       0       0       1       0    6.6707663       789
          5    4       7       1       0       1       0    6.0799332       437
          5    5       5       0       0       0       1    7.0535857      1157
          5    6      12       1       0       0       1    7.6783264      2161
          5    7       0       0       0       0       0         .            0
          5    8       1       1       0       0       0    6.295266        542
```

5.2.3 Progabide data

The Progabide data have been analyzed in many places; they are available in Thall and Vail (1990). The data are from a panel study in which four successive two-week counts of seizures were recorded for each epileptic patient in the study. The covariates are the Progabide treatment indicator (0=placebo, 1=Progabide), the followup indicator (0=baseline measurement, 1=followup), and an interaction of these covariates.

We use only these covariates. Other sources that analyze these data include additional covariates including the age of the patient.

A description of the panel structure of the data is:

```
        id:  1, 2, ..., 59                              n =         59
         t:  0, 1, ..., 4                               T =          5
             Delta(t) = 1; (4-0)+1 = 5
             (id*t uniquely identifies each observation)

Distribution of T_i:    min       5%      25%     50%     75%     95%     max
                          5        5        5       5       5       5       5

       Freq.  Percent   Cum. |  Pattern
       ---------------------------+---------
          59   100.00  100.00 |  11111
       ---------------------------+---------
          59   100.00         |  XXXXX
```

A summary of the variables is:

```
Variable         |      Mean   Std. Dev.       Min        Max |    Observations
-----------------+--------------------------------------------+----------------
seizures overall|   12.86441    18.68797         0        151 |    N =     295
         between|               13.90844       2.2       90.6 |    n =      59
         within |               12.58679 -14.73559   77.06441 |    T =       5
                 |                                             |
time     overall|         .8    .4006797         0          1 |    N =     295
         between|                      0        .8         .8 |    n =      59
         within |               .4006797         0          1 |    T =       5
                 |                                             |
progabide overall|   .5254237    .5002017         0          1 |    N =     295
         between|               .5036396         0          1 |    n =      59
         within |                      0  .5254237   .5254237 |    T =       5
```

timeXprog	overall	.420339	.4944521	0	1	N =	295
	between		.4029117	0	.8	n =	59
	within		.2904372	-.379661	.620339	T =	5
lnPeriod	overall	.9704061	.55546	.6931472	2.079442	N =	295
	between		0	.9704061	.9704061	n =	59
	within		.55546	.6931472	2.079442	T =	5

The data are:

id	t	seizures	time	progabide	timeXprog	lnPeriod
1	0	11	0	0	0	2.079442
1	1	5	1	0	0	.6931472
1	2	3	1	0	0	.6931472
1	3	3	1	0	0	.6931472
1	4	3	1	0	0	.6931472
2	0	11	0	0	0	2.079442
2	1	3	1	0	0	.6931472
2	2	5	1	0	0	.6931472
2	3	3	1	0	0	.6931472
2	4	3	1	0	0	.6931472
3	0	6	0	0	0	2.079442
3	1	2	1	0	0	.6931472
3	2	4	1	0	0	.6931472
3	3	0	1	0	0	.6931472
3	4	5	1	0	0	.6931472
4	0	8	0	0	0	2.079442
4	1	4	1	0	0	.6931472
4	2	4	1	0	0	.6931472
4	3	1	1	0	0	.6931472
4	4	4	1	0	0	.6931472
5	0	66	0	0	0	2.079442
5	1	7	1	0	0	.6931472
5	2	18	1	0	0	.6931472
5	3	9	1	0	0	.6931472
5	4	21	1	0	0	.6931472
6	0	27	0	0	0	2.079442
6	1	5	1	0	0	.6931472
6	2	2	1	0	0	.6931472
6	3	8	1	0	0	.6931472
6	4	7	1	0	0	.6931472
7	0	12	0	0	0	2.079442
7	1	6	1	0	0	.6931472
7	2	4	1	0	0	.6931472
7	3	0	1	0	0	.6931472
7	4	2	1	0	0	.6931472
8	0	52	0	0	0	2.079442
8	1	40	1	0	0	.6931472
8	2	20	1	0	0	.6931472
8	3	23	1	0	0	.6931472
8	4	12	1	0	0	.6931472
9	0	23	0	0	0	2.079442
9	1	5	1	0	0	.6931472
9	2	6	1	0	0	.6931472
9	3	6	1	0	0	.6931472
9	4	5	1	0	0	.6931472
10	0	10	0	0	0	2.079442
10	1	14	1	0	0	.6931472
10	2	13	1	0	0	.6931472

10	3	6	1	0	0	.6931472
10	4	0	1	0	0	.6931472
11	0	52	0	0	0	2.079442
11	1	26	1	0	0	.6931472
11	2	12	1	0	0	.6931472
11	3	6	1	0	0	.6931472
11	4	22	1	0	0	.6931472
12	0	33	0	0	0	2.079442
12	1	12	1	0	0	.6931472
12	2	6	1	0	0	.6931472
12	3	8	1	0	0	.6931472
12	4	5	1	0	0	.6931472
13	0	18	0	0	0	2.079442
13	1	4	1	0	0	.6931472
13	2	4	1	0	0	.6931472
13	3	6	1	0	0	.6931472
13	4	2	1	0	0	.6931472
14	0	42	0	0	0	2.079442
14	1	7	1	0	0	.6931472
14	2	9	1	0	0	.6931472
14	3	12	1	0	0	.6931472
14	4	14	1	0	0	.6931472
15	0	87	0	0	0	2.079442
15	1	16	1	0	0	.6931472
15	2	24	1	0	0	.6931472
15	3	10	1	0	0	.6931472
15	4	9	1	0	0	.6931472
16	0	50	0	0	0	2.079442
16	1	11	1	0	0	.6931472
16	2	0	1	0	0	.6931472
16	3	0	1	0	0	.6931472
16	4	5	1	0	0	.6931472
17	0	18	0	0	0	2.079442
17	1	0	1	0	0	.6931472
17	2	0	1	0	0	.6931472
17	3	3	1	0	0	.6931472
17	4	3	1	0	0	.6931472
18	0	111	0	0	0	2.079442
18	1	37	1	0	0	.6931472
18	2	29	1	0	0	.6931472
18	3	28	1	0	0	.6931472
18	4	29	1	0	0	.6931472
19	0	18	0	0	0	2.079442
19	1	3	1	0	0	.6931472
19	2	5	1	0	0	.6931472
19	3	2	1	0	0	.6931472
19	4	5	1	0	0	.6931472
20	0	20	0	0	0	2.079442
20	1	3	1	0	0	.6931472
20	2	0	1	0	0	.6931472
20	3	6	1	0	0	.6931472
20	4	7	1	0	0	.6931472
21	0	12	0	0	0	2.079442
21	1	3	1	0	0	.6931472
21	2	4	1	0	0	.6931472
21	3	3	1	0	0	.6931472
21	4	4	1	0	0	.6931472
22	0	9	0	0	0	2.079442
22	1	3	1	0	0	.6931472
22	2	4	1	0	0	.6931472
22	3	3	1	0	0	.6931472

DATASETS

22	4	4	1	0	0	.6931472
23	0	17	0	0	0	2.079442
23	1	2	1	0	0	.6931472
23	2	3	1	0	0	.6931472
23	3	3	1	0	0	.6931472
23	4	5	1	0	0	.6931472
24	0	28	0	0	0	2.079442
24	1	8	1	0	0	.6931472
24	2	12	1	0	0	.6931472
24	3	2	1	0	0	.6931472
24	4	8	1	0	0	.6931472
25	0	55	0	0	0	2.079442
25	1	18	1	0	0	.6931472
25	2	24	1	0	0	.6931472
25	3	76	1	0	0	.6931472
25	4	25	1	0	0	.6931472
26	0	9	0	0	0	2.079442
26	1	2	1	0	0	.6931472
26	2	1	1	0	0	.6931472
26	3	2	1	0	0	.6931472
26	4	1	1	0	0	.6931472
27	0	10	0	0	0	2.079442
27	1	3	1	0	0	.6931472
27	2	1	1	0	0	.6931472
27	3	4	1	0	0	.6931472
27	4	2	1	0	0	.6931472
28	0	47	0	0	0	2.079442
28	1	13	1	0	0	.6931472
28	2	15	1	0	0	.6931472
28	3	13	1	0	0	.6931472
28	4	12	1	0	0	.6931472
29	0	76	0	1	0	2.079442
29	1	11	1	1	1	.6931472
29	2	14	1	1	1	.6931472
29	3	9	1	1	1	.6931472
29	4	8	1	1	1	.6931472
30	0	38	0	1	0	2.079442
30	1	8	1	1	1	.6931472
30	2	7	1	1	1	.6931472
30	3	9	1	1	1	.6931472
30	4	4	1	1	1	.6931472
31	0	19	0	1	0	2.079442
31	1	0	1	1	1	.6931472
31	2	4	1	1	1	.6931472
31	3	3	1	1	1	.6931472
31	4	0	1	1	1	.6931472
32	0	10	0	1	0	2.079442
32	1	3	1	1	1	.6931472
32	2	6	1	1	1	.6931472
32	3	1	1	1	1	.6931472
32	4	3	1	1	1	.6931472
33	0	19	0	1	0	2.079442
33	1	2	1	1	1	.6931472
33	2	6	1	1	1	.6931472
33	3	7	1	1	1	.6931472
33	4	4	1	1	1	.6931472
34	0	24	0	1	0	2.079442
34	1	4	1	1	1	.6931472
34	2	3	1	1	1	.6931472
34	3	1	1	1	1	.6931472
34	4	3	1	1	1	.6931472

35	0	31	0	1	0	2.079442
35	1	22	1	1	1	.6931472
35	2	17	1	1	1	.6931472
35	3	19	1	1	1	.6931472
35	4	16	1	1	1	.6931472
36	0	14	0	1	0	2.079442
36	1	5	1	1	1	.6931472
36	2	4	1	1	1	.6931472
36	3	7	1	1	1	.6931472
36	4	4	1	1	1	.6931472
37	0	11	0	1	0	2.079442
37	1	2	1	1	1	.6931472
37	2	4	1	1	1	.6931472
37	3	0	1	1	1	.6931472
37	4	4	1	1	1	.6931472
38	0	67	0	1	0	2.079442
38	1	3	1	1	1	.6931472
38	2	7	1	1	1	.6931472
38	3	7	1	1	1	.6931472
38	4	7	1	1	1	.6931472
39	0	41	0	1	0	2.079442
39	1	4	1	1	1	.6931472
39	2	18	1	1	1	.6931472
39	3	2	1	1	1	.6931472
39	4	5	1	1	1	.6931472
40	0	7	0	1	0	2.079442
40	1	2	1	1	1	.6931472
40	2	1	1	1	1	.6931472
40	3	1	1	1	1	.6931472
40	4	0	1	1	1	.6931472
41	0	22	0	1	0	2.079442
41	1	0	1	1	1	.6931472
41	2	2	1	1	1	.6931472
41	3	4	1	1	1	.6931472
41	4	0	1	1	1	.6931472
42	0	13	0	1	0	2.079442
42	1	5	1	1	1	.6931472
42	2	4	1	1	1	.6931472
42	3	0	1	1	1	.6931472
42	4	3	1	1	1	.6931472
43	0	46	0	1	0	2.079442
43	1	11	1	1	1	.6931472
43	2	14	1	1	1	.6931472
43	3	25	1	1	1	.6931472
43	4	15	1	1	1	.6931472
44	0	36	0	1	0	2.079442
44	1	10	1	1	1	.6931472
44	2	5	1	1	1	.6931472
44	3	3	1	1	1	.6931472
44	4	8	1	1	1	.6931472
45	0	38	0	1	0	2.079442
45	1	19	1	1	1	.6931472
45	2	7	1	1	1	.6931472
45	3	6	1	1	1	.6931472
45	4	7	1	1	1	.6931472
46	0	7	0	1	0	2.079442
46	1	1	1	1	1	.6931472
46	2	1	1	1	1	.6931472
46	3	2	1	1	1	.6931472
46	4	4	1	1	1	.6931472
47	0	36	0	1	0	2.079442

DATASETS

47	1	6	1	1	1	.6931472
47	2	10	1	1	1	.6931472
47	3	8	1	1	1	.6931472
47	4	8	1	1	1	.6931472
48	0	11	0	1	0	2.079442
48	1	2	1	1	1	.6931472
48	2	1	1	1	1	.6931472
48	3	0	1	1	1	.6931472
48	4	0	1	1	1	.6931472
49	0	151	0	1	0	2.079442
49	1	102	1	1	1	.6931472
49	2	65	1	1	1	.6931472
49	3	72	1	1	1	.6931472
49	4	63	1	1	1	.6931472
50	0	22	0	1	0	2.079442
50	1	4	1	1	1	.6931472
50	2	3	1	1	1	.6931472
50	3	2	1	1	1	.6931472
50	4	4	1	1	1	.6931472
51	0	42	0	1	0	2.079442
51	1	8	1	1	1	.6931472
51	2	6	1	1	1	.6931472
51	3	5	1	1	1	.6931472
51	4	7	1	1	1	.6931472
52	0	32	0	1	0	2.079442
52	1	1	1	1	1	.6931472
52	2	3	1	1	1	.6931472
52	3	1	1	1	1	.6931472
52	4	5	1	1	1	.6931472
53	0	56	0	1	0	2.079442
53	1	18	1	1	1	.6931472
53	2	11	1	1	1	.6931472
53	3	28	1	1	1	.6931472
53	4	13	1	1	1	.6931472
54	0	24	0	1	0	2.079442
54	1	6	1	1	1	.6931472
54	2	3	1	1	1	.6931472
54	3	4	1	1	1	.6931472
54	4	0	1	1	1	.6931472
55	0	16	0	1	0	2.079442
55	1	3	1	1	1	.6931472
55	2	5	1	1	1	.6931472
55	3	4	1	1	1	.6931472
55	4	3	1	1	1	.6931472
56	0	22	0	1	0	2.079442
56	1	1	1	1	1	.6931472
56	2	23	1	1	1	.6931472
56	3	19	1	1	1	.6931472
56	4	8	1	1	1	.6931472
57	0	25	0	1	0	2.079442
57	1	2	1	1	1	.6931472
57	2	3	1	1	1	.6931472
57	3	0	1	1	1	.6931472
57	4	1	1	1	1	.6931472
58	0	13	0	1	0	2.079442
58	1	0	1	1	1	.6931472
58	2	0	1	1	1	.6931472
58	3	0	1	1	1	.6931472
58	4	0	1	1	1	.6931472
59	0	12	0	1	0	2.079442
59	1	1	1	1	1	.6931472

```
                59          2          4          1          1          1      .6931472
                59          3          3          1          1          1      .6931472
                59          4          2          1          1          1      .6931472
```

5.2.4 Simulated logistic data

This dataset is used in Chapter 4 to illustrate the calculation and interpretation of the QIC diagnostic measure. The panel identifier is given by id; the repeated measures for the panel are given by t; the x1 covariate measures are random $U(0,1)$; the x2 covariate measures are random $N(0,1)$; the x3 covariate measures are random $U(5,10)$; and the binary outcome y is generated such that .4 is the approximate within-panel correlation and

$$y_{it} = -.4\text{x}1 + .25\text{x}2 + .15\text{x}3 + 1.3 \tag{5.1}$$

for a marginal logistic regression.

A description of the panel structure of the data is:

```
       id:  1, 2, ..., 50                                      n =           50
        t:  1, 2, ..., 8                                       T =            8
            Delta(t) = 1; (8-1)+1 = 8
            (id*t uniquely identifies each observation)

Distribution of T_i:     min       5%        25%       50%       75%       95%       max
                           8        8         8         8         8         8         8

        Freq.   Percent     Cum. |  Pattern
        ------------------------+----------
          50    100.00    100.00 |  11111111
        ------------------------+----------
          50    100.00           |  XXXXXXXX
```

A summary of the variables is:

```
Variable      |      Mean    Std. Dev.       Min        Max |   Observations
--------------+---------------------------------------------+----------------
y    overall  |       .74    .4391836          0          1 |   N =     400
     between  |            .2866503           0          1 |   n =      50
     within   |            .3348961       -.135      1.615 |   T =       8
              |                                              |
x1   overall  |   .4873817   .2843164    .0022786   .9914105 |   N =     400
     between  |            .0966579    .2572445    .690023  |   n =      50
     within   |            .2676882   -.178456   1.107685  |   T =       8
              |                                              |
x2   overall  |  -.0616686   .9793633  -2.391706   3.070278 |   N =     400
     between  |            .3630255   -.7829833   .6286468 |   n =      50
     within   |            .9108661  -2.282303   2.729702  |   T =       8
              |                                              |
x3   overall  |   .0904452   2.778632  -4.998804   4.942484 |   N =     400
     between  |            .9949208   -1.864617  1.897457  |   n =      50
     within   |            2.597748  -5.857181   6.516726  |   T =       8
```

DATASETS

The data are:

id	t	y	x1	x2	x3
1	1	1	.88523501	-.26417433	4.3258062
1	2	1	.33954228	-1.4430173	3.4140513
1	3	1	.98692055	-.16949629	.60771281
1	4	1	.43864285	-.12070738	4.3190559
1	5	1	.19520924	.51354993	-.76819588
1	6	1	.62452864	.61329929	-1.8277712
1	7	1	.58458474	-.29184848	4.3657006
1	8	1	.40398331	1.2441504	-.08235102
2	1	1	.5432351	-.39363134	1.0439372
2	2	1	.45269737	-.65273565	-1.547725
2	3	1	.52367102	-.9039997	4.5063062
2	4	1	.7373026	-.7703287	.76219818
2	5	1	.84405544	.31304621	3.7118576
2	6	1	.26029491	.1485929	-.42186429
2	7	1	.68926677	-1.1141923	-4.1421549
2	8	1	.33319191	.64182082	1.740416
3	1	1	.99141046	.14224213	.17241989
3	2	1	.11734212	1.995702	3.0959682
3	3	1	.17001171	.37615613	1.4662582
3	4	1	.95515468	-.36269575	1.8412877
3	5	1	.26147315	-1.1041402	4.4556939
3	6	1	.29333199	-.79240046	-1.7109089
3	7	1	.62018647	-.0337262	-4.7023703
3	8	1	.87830916	2.4218153	-4.5688076
4	1	1	.6308325	.03634451	3.4425292
4	2	1	.17430338	.72096278	1.4317767
4	3	1	.43297697	.61541613	2.3123194
4	4	1	.44872869	1.0482384	2.1555201
4	5	1	.53018721	1.5200186	.46990644
4	6	1	.88605676	.42986316	.98008882
4	7	1	.58982214	.6308999	4.7580965
4	8	1	.1641364	-.99606258	-1.8691156
5	1	1	.66959327	.56058246	.91035143
5	2	1	.93625849	.55995908	3.6458639
5	3	0	.76221961	-1.2894494	3.4636962
5	4	0	.87525675	-1.0410228	-1.5899535
5	5	0	.40668204	-.37620679	-2.0461566
5	6	1	.53499123	.06020311	3.1828135
5	7	1	.32165161	-.30549954	-3.2824211
5	8	0	.44859654	-1.1548636	.11750016
6	1	1	.89396044	-.20606491	-2.6077053
6	2	1	.95474025	.27077224	-4.9657403
6	3	1	.44044753	.20482204	-3.2177994
6	4	1	.94513756	-1.053525	3.1326672
6	5	1	.40861741	1.2710332	.59599184
6	6	0	.61684079	-.25491999	1.3331989
6	7	1	.43013996	-.32949256	1.7022658
6	8	1	.28627501	-.33010354	-.81824249
7	1	0	.75287802	-.47702251	-.3125868
7	2	1	.94853001	.38926723	3.5482391
7	3	1	.02615721	-1.2221744	1.0403809
7	4	0	.42648522	-.86363652	-2.2260864
7	5	0	.01091767	-.87978583	-3.3014733
7	6	0	.55131528	-.19303872	-4.5931443
7	7	0	.39179926	-.87915599	-2.3541865
7	8	0	.35669087	-.40892197	3.6668923
8	1	1	.12855085	.30357532	3.0231191
8	2	0	.2241307	-.00238475	-1.463341

8	3	1	.06231786	.60146353	-4.9988038
8	4	1	.47453826	1.5588749	.54277238
8	5	0	.48496416	-.18722218	-3.8340121
8	6	0	.77902541	-.52048368	-2.0633239
8	7	1	.53028459	-.1661389	-4.3329232
8	8	1	.80609256	.46771013	.96578931
9	1	1	.88378248	.63588368	2.226928
9	2	1	.15312962	-.38212096	1.8735343
9	3	0	.90788475	-1.286626	-3.236961
9	4	0	.834628	-.13891725	-1.3035694
9	5	1	.3549188	-.65583187	.44230726
9	6	0	.61216385	.0449998	-.70483647
9	7	1	.79174163	-1.3466392	3.7028283
9	8	0	.83288793	1.2581238	-3.6604588
10	1	1	.34653541	-.64545144	-.72763472
10	2	1	.5918418	-1.1211371	-1.6468374
10	3	1	.82795863	-1.4299645	.9285548
10	4	1	.20497884	.50605129	-4.9833893
10	5	1	.53749413	-.51641011	-.71150266
10	6	1	.50569005	-.63162137	1.7601965
10	7	1	.50079654	-.36714636	-1.3482761
10	8	1	.3367656	-.58090701	-3.9945315
11	1	1	.00380824	-1.9394894	3.5328471
11	2	1	.89680016	.88590003	-.47946425
11	3	1	.1360347	-1.0620764	1.1742915
11	4	1	.69414398	-.0589215	1.8255025
11	5	1	.38013169	1.6410543	.71103245
11	6	1	.1684285	-1.1057832	2.921645
11	7	1	.31684049	.28738969	-4.8851947
11	8	0	.04945258	-.84357201	-3.0051272
12	1	0	.63517814	-.44529748	3.4268863
12	2	1	.76923942	-1.5517598	1.7410977
12	3	1	.76285548	.00949782	2.1936293
12	4	1	.58721832	-1.2622061	1.1906233
12	5	1	.91543622	.07091063	-4.3739526
12	6	0	.0366871	-1.312804	.08243607
12	7	1	.29761236	-.25571016	-2.9577733
12	8	0	.02741682	-1.0394488	-2.044473
13	1	0	.53181705	-.1171302	-3.4598432
13	2	1	.51958141	-.44810463	.89557035
13	3	0	.60291944	.79294672	-1.8818473
13	4	0	.02950277	-.14219833	1.1538704
13	5	0	.32117461	.22017125	-2.3138548
13	6	1	.25505851	.4592555	-1.0209359
13	7	0	.30665354	1.8549507	1.4762511
13	8	1	.4827527	.39750888	4.675657
14	1	1	.56189396	-.13241135	-.12640056
14	2	1	.5073453	-.2808839	2.1322419
14	3	1	.62189489	.15450659	4.654006
14	4	1	.04573518	.28300012	-2.4122847
14	5	1	.83306196	-.13693423	2.9582598
14	6	1	.95508826	1.0661567	3.911096
14	7	1	.8425308	.17461357	-.02330345
14	8	1	.22936638	-1.4735188	3.7814739
15	1	1	.07859847	-1.098752	-3.6369593
15	2	1	.66571564	-.4837278	.50674996
15	3	1	.34542704	-2.1701744	-4.1409653
15	4	1	.38745922	-.48917295	-.3742053
15	5	1	.98387702	.88562335	-4.0334277
15	6	0	.86916942	-.00397613	-.19389177
15	7	1	.74294057	.0916879	-3.2145793

15	8	1	.27558712	.34792129	3.430073
16	1	1	.58397944	-.91380405	-4.425422
16	2	1	.5126295	.08168254	1.6236404
16	3	1	.41536157	-1.8501342	-2.1072937
16	4	0	.48953521	1.2301491	-2.9883699
16	5	1	.68019157	-.28588617	-1.5320506
16	6	1	.3130354	.07511148	-.90140523
16	7	1	.12564593	.61939931	.14262375
16	8	1	.11187084	-.4230035	2.4722555
17	1	0	.37782425	-.02957512	-3.6807588
17	2	0	.00227859	-.3834609	-1.0739004
17	3	1	.07346247	1.3034031	.70664593
17	4	0	.18263109	1.2907545	-3.3018123
17	5	1	.36099134	-.26646035	3.957581
17	6	1	.57799378	1.1626149	-.90230217
17	7	0	.72386528	-.5290824	-.17979672
17	8	1	.98178964	-.43350923	-.6303771
18	1	1	.52735318	-.13539128	.24123824
18	2	1	.9006656	-.07789571	-1.3711217
18	3	1	.24406173	-.78731442	-4.9034452
18	4	1	.25782954	-.07930608	-.5210331
18	5	1	.62259568	.66037158	4.5491875
18	6	1	.98445967	-.37126614	-2.9832802
18	7	1	.41295201	-.74630656	-3.561139
18	8	1	.17591862	.80575665	3.5713269
19	1	1	.29881176	.44737735	-.07935133
19	2	1	.67180651	2.0835332	-4.1586076
19	3	1	.71226042	.99170182	4.3754299
19	4	1	.67633982	1.3345701	-4.6777002
19	5	1	.92177484	.31909905	2.2397399
19	6	1	.6557825	-.60688805	3.9729033
19	7	1	.77172166	-.50578954	-.63316955
19	8	1	.26164491	-.97731943	-.74951374
20	1	1	.17978976	-.89168252	-4.995013
20	2	1	.00593496	.93206089	2.192063
20	3	1	.34522656	2.3692898	4.5985832
20	4	1	.81558473	-1.2878287	-3.5325098
20	5	1	.22084109	1.3172106	3.6624955
20	6	1	.22060169	.70023263	.47788956
20	7	1	.8652202	-.00889777	-1.0808749
20	8	1	.51823612	.53777187	3.1110475
21	1	1	.13201322	-1.3715155	1.9826525
21	2	0	.7645131	-.21982861	4.1855778
21	3	0	.69539795	-1.8811427	4.3522186
21	4	0	.2077739	-1.1611297	.77274462
21	5	0	.73578102	-.12590762	-1.5178568
21	6	0	.19837309	-.98063133	-4.8624581
21	7	0	.48287501	-.04722225	.67162919
21	8	1	.20858556	-.47648833	3.0968339
22	1	0	.53156268	-.4926744	-4.8648589
22	2	1	.2012225	-.64922471	2.5147342
22	3	1	.02760816	-1.5883851	3.5598681
22	4	0	.16596069	-2.0535783	-4.2191406
22	5	0	.68825371	-.35866027	-1.9256883
22	6	1	.84784981	.27945389	4.9424842
22	7	1	.86893977	.68770709	2.6221845
22	8	1	.16075925	-1.8818703	-1.5203959
23	1	0	.83635474	-.69712701	-1.9354771
23	2	1	.04011063	-.09210309	2.8352676
23	3	1	.40751043	.34407518	-.77299453
23	4	0	.8326156	-.12476642	-3.6740776

23	5	1	.68758874	1.2680465	2.212705
23	6	0	.27736921	-1.7895554	3.2003125
23	7	1	.4968114	.19404024	3.2960769
23	8	1	.22005071	-1.2299077	4.7701399
24	1	0	.23589676	-1.3315594	-1.7132597
24	2	0	.17549448	-.54896919	3.9754857
24	3	1	.19661888	-.74119281	-1.3661813
24	4	1	.2569721	-.73098404	.58565707
24	5	1	.66256802	-.33598341	-1.6845021
24	6	1	.34392516	-.02288244	.66320471
24	7	1	.24420482	.45729008	-1.0345737
24	8	1	.65872017	-.41720822	-1.6707589
25	1	1	.23582646	.84608032	4.4993701
25	2	1	.77706614	-.16192432	-.30763881
25	3	1	.5493553	-1.5005625	2.9079465
25	4	1	.18284525	.75113585	-4.1398297
25	5	1	.33852097	-.05585918	-.99301879
25	6	1	.26992076	-.43186582	-3.0121238
25	7	1	.98637592	.13750156	-3.6957577
25	8	1	.30449762	-.19142843	2.2392974
26	1	1	.96272796	1.4228575	-4.6142312
26	2	1	.02418522	-.32378812	-1.2151424
26	3	1	.92729395	-.64441027	.19044551
26	4	1	.89722758	1.0411978	2.4677298
26	5	1	.72422163	.64399618	2.3188249
26	6	1	.69992838	.34498084	-.99608916
26	7	1	.62149437	1.1166724	4.7785935
26	8	1	.6631046	-.63930528	-4.7249802
27	1	1	.74619942	1.2856775	-3.0055967
27	2	1	.29762063	.2264596	-3.331311
27	3	1	.58929351	1.3225808	-1.6281139
27	4	1	.37908853	-1.1701139	-4.3093914
27	5	1	.19380906	1.0166835	-2.0458075
27	6	1	.29698769	-.28090275	-1.2872821
27	7	0	.22582572	-.56469194	-3.6555851
27	8	1	.82906298	-1.0804181	4.5924508
28	1	1	.32449638	-.66505316	1.9937702
28	2	1	.62007629	1.6773657	2.9524903
28	3	1	.39488068	.07470998	-1.3430214
28	4	0	.63217978	.48602902	4.2420256
28	5	1	.33537316	.8967287	.05658468
28	6	1	.21462002	.35152923	1.1532962
28	7	1	.71845174	-.68620416	1.1975402
28	8	1	.49786685	2.0024899	-1.7361389
29	1	0	.58656109	.39099986	3.2369623
29	2	1	.8677033	.02217869	-3.3658257
29	3	0	.85653838	3.0142648	-4.5343759
29	4	1	.3687826	-1.6364854	2.6264919
29	5	1	.65798499	.69043478	3.2047573
29	6	1	.17775392	1.6158611	-.190484
29	7	1	.91381994	-.17701318	-.35774313
29	8	0	.9393249	-1.1346761	-2.3318768
30	1	1	.64733838	-.51795779	.43073647
30	2	1	.4630548	.74890016	-.97820477
30	3	1	.53684018	-.77432798	3.3561943
30	4	1	.37143225	1.3865299	.35617686
30	5	1	.60971577	-1.0785653	.66651807
30	6	1	.56700373	.08187213	-4.3671494
30	7	0	.95745034	.65348742	3.067481
30	8	1	.15581061	-.9164123	.51913779
31	1	1	.40739452	3.070278	.53683195

DATASETS

31	2	1	.56327384	-.3849021	-.70206121
31	3	1	.29046376	.60229091	-3.7232115
31	4	1	.53421342	.14574362	2.0436923
31	5	1	.54962087	-1.9203826	2.2235067
31	6	1	.95600868	1.6891462	3.6981974
31	7	1	.63389869	-.78329977	3.0074837
31	8	1	.64621045	-.18761821	3.798925
32	1	0	.10232071	.45032416	-4.937239
32	2	0	.78922367	1.2443122	1.1670747
32	3	0	.42552667	.54744884	2.3350909
32	4	0	.41535373	-.69373723	-3.1032691
32	5	0	.763188	-.5946792	-.5058964
32	6	0	.72860125	-.01476781	-4.9182028
32	7	0	.13680069	.16688668	-4.1165028
32	8	0	.91027842	.55757484	-.8379931
33	1	1	.54590047	.17411157	-.21226072
33	2	1	.83071607	-.73925464	-2.1004965
33	3	1	.17521875	-.46675134	3.02218
33	4	1	.61075648	-.8507672	1.6696662
33	5	1	.01699386	-1.2641486	4.4708352
33	6	1	.70993419	.3947986	-2.4209945
33	7	1	.06030223	1.7805091	.54445033
33	8	1	.06520496	-.96107412	-4.7515674
34	1	1	.86807731	-.10076391	-4.7624085
34	2	1	.21507805	1.5387058	-1.4352469
34	3	1	.76262093	.60498603	-.36165949
34	4	1	.80968973	1.2913217	3.6508464
34	5	1	.25898625	-1.3883636	-2.6991054
34	6	1	.32008237	-.36760313	.42779067
34	7	1	.64347544	.32773953	-1.753068
34	8	1	.03699922	-1.8895956	-2.1669865
35	1	1	.78041924	.70382262	-1.5137079
35	2	1	.65758922	.88996758	-.17061118
35	3	1	.34141978	-1.2497238	2.5472233
35	4	1	.39119715	-.41529182	-2.1858799
35	5	1	.29317256	-.63752937	.43308025
35	6	1	.3642809	-.57384934	-1.2935386
35	7	1	.24336802	1.3593231	4.0431023
35	8	1	.340014	-.70663325	2.7504995
36	1	1	.13895188	-.15480263	.47988757
36	2	0	.53017192	-2.3917064	.32486999
36	3	1	.1484553	-2.2279841	-3.2921182
36	4	0	.03120775	2.4427257	-3.2128021
36	5	1	.19390937	1.892197	-1.2066829
36	6	1	.93600489	-.38750505	-2.369094
36	7	1	.51341005	-.01496864	3.5563123
36	8	1	.60332375	-.52653447	-2.6556237
37	1	0	.80361478	.21759436	4.0917264
37	2	1	.83905361	-.202704	-.1131667
37	3	1	.66134728	.53206591	.05741577
37	4	1	.14588421	1.0827364	-2.3197276
37	5	0	.98844257	-.46330086	-2.9275867
37	6	0	.55946653	-.19740824	1.0410657
37	7	1	.77506609	-1.332941	4.4338478
37	8	1	.47633007	-1.1817738	-.73263268
38	1	1	.25587877	-.81091915	3.646419
38	2	1	.31655058	1.0162053	.18708434
38	3	1	.47828454	.09232223	-2.7465362
38	4	1	.5520271	.73310513	-.7492126
38	5	1	.70153985	1.1413866	1.1203905
38	6	1	.08902159	-1.2584207	4.7620969

38	7	1	.94034435	.41596148	4.1898823
38	8	1	.59914863	1.1058064	2.8142233
39	1	1	.56260686	2.1370153	2.0741688
39	2	1	.23018233	.30045069	3.0684806
39	3	1	.04914308	-.08990481	-.97440486
39	4	1	.01188321	-1.2300802	.45396775
39	5	1	.38878313	-.56517104	1.5995643
39	6	1	.34094476	-.96707895	-.15632532
39	7	1	.67529112	.67396387	-.01713385
39	8	1	.69555857	-.16860392	.42847959
40	1	1	.19676667	-1.2299737	3.6654478
40	2	1	.89083925	.51698281	1.5202862
40	3	1	.02532814	.87571896	-.52499704
40	4	1	.69018445	-1.019677	-.77943381
40	5	1	.2821059	-.05965368	-3.4306597
40	6	1	.85792841	-.12587525	-3.124772
40	7	1	.73548534	-1.0268592	-4.3759957
40	8	1	.42788794	-.52243421	-.77225534
41	1	1	.08003791	.94269499	2.2493524
41	2	1	.16347113	-.18662671	4.2082316
41	3	1	.83242888	.31693476	-1.1179924
41	4	1	.21015557	-.73428691	1.9414704
41	5	1	.8732753	-2.0275832	-.95547237
41	6	0	.28980839	.05248421	2.5167067
41	7	1	.71055052	-.7547069	-4.7631615
41	8	1	.52896684	-1.1094163	4.5933701
42	1	0	.12264861	.22211133	1.9060186
42	2	0	.68375949	-1.7287251	2.0987929
42	3	1	.16295932	-.88180576	-1.7087712
42	4	0	.65459204	.28438627	-.98221578
42	5	0	.93183955	.74176345	2.3216357
42	6	0	.0145188	-1.3564762	-2.2650536
42	7	0	.9675157	-1.833475	2.6520978
42	8	1	.44257704	.90140632	2.2830742
43	1	1	.13219481	-.0285642	2.3297012
43	2	1	.43382877	2.3835479	3.823715
43	3	1	.80208312	1.5250206	-1.5980395
43	4	1	.18421517	.07678258	-.00703386
43	5	0	.54542541	-1.2414406	-3.0611441
43	6	0	.03521744	-1.75639	-1.041176
43	7	1	.70278961	-.3389013	2.0592166
43	8	1	.75151	.84971109	-3.3548854
44	1	1	.8386282	2.6490389	4.1333626
44	2	1	.83382061	-.73016641	-2.0150525
44	3	1	.36430141	.49163107	.54000746
44	4	1	.84947008	.90074762	-2.1433894
44	5	0	.88426841	-.85493851	-1.9446169
44	6	1	.33416062	.92205232	-.37553885
44	7	0	.83010641	.52395781	4.6679322
44	8	1	.02928144	1.126852	3.3548694
45	1	0	.38704076	-.02501493	-3.9180342
45	2	0	.85878778	.34433422	-4.4539489
45	3	1	.36038781	-.00653701	4.202635
45	4	0	.88836044	-.92589448	-3.9760112
45	5	0	.17073028	.48080668	-4.058332
45	6	0	.84309061	-2.0786798	.61482275
45	7	0	.31751812	-.85146016	3.1539063
45	8	0	.26018423	.80880252	.08828484
46	1	0	.81664001	2.3827927	1.2865451
46	2	0	.18930089	-.18557107	-1.107921
46	3	0	.14386472	1.055132	1.1477528

46	4	0	.05711375	-1.8183438	-2.3616186
46	5	1	.25490037	1.2199273	1.6594807
46	6	0	.88611686	-1.4254149	-3.1099571
46	7	1	.93178333	1.8158332	2.2550158
46	8	1	.76665744	-.65296269	4.6473839
47	1	1	.54213497	2.2005336	.28687697
47	2	1	.31808571	-1.1984512	1.3526689
47	3	1	.15070568	-.42601789	4.7049423
47	4	1	.85729343	1.1353603	1.1759126
47	5	0	.59699909	-1.663961	-3.2895913
47	6	1	.07067903	.44522875	-3.4154149
47	7	1	.77577306	.04262007	-4.0131043
47	8	1	.09566843	1.591931	-.65167378
48	1	0	.4154052	1.2803466	3.7479917
48	2	0	.30757364	-.53293749	3.5696961
48	3	0	.14002295	.21588168	2.73561
48	4	1	.16449026	.8598506	4.1874599
48	5	0	.04353584	.72386736	-.88665317
48	6	1	.03160888	-.66081645	2.5444921
48	7	0	.87754801	.01414242	-2.0073164
48	8	0	.07777111	-1.2705579	1.2883756
49	1	0	.55738669	-.80888332	-2.8039781
49	2	0	.43946018	.03326704	-2.9146663
49	3	1	.86576495	-.31548117	4.343178
49	4	0	.49062378	-.05854831	-1.9599561
49	5	0	.89837667	-.49643377	-4.4098143
49	6	0	.05956974	-.38045395	3.1175952
49	7	1	.1071656	-.94569154	3.7714548
49	8	0	.28174196	-.31755863	-3.0976467
50	1	0	.48683077	.41736385	4.0366273
50	2	1	.16975154	.70538924	1.677822
50	3	1	.80913531	-1.3701565	.97360708
50	4	1	.09923071	.36496528	2.2439137
50	5	1	.51029362	-.02965048	1.7768015
50	6	1	.24087059	-.6967609	-1.0320388
50	7	0	.37647759	-2.0212218	-.89256952
50	8	1	.03292408	-.74457299	.56396776

5.2.5 Simulated user-specified correlated data

This dataset is used in Chapter 3 to illustrate calculation of a PA-GEE model with user-specified correlation structure. The panel identifier is given by id; the repeated measures for the panel are given by t; the x1 covariate measures are random binary; the x2 covariate measures are random binary; and the binary outcome y is generated such that

$$y_{it} = x1 + x2 + 1 \tag{5.2}$$

for a marginal linear regression with theoretical panel correlation given by

$$\mathbf{R} = \begin{bmatrix} 1 & .6 & 0 & 0 & 0 & 0 & 0 & 0 \\ .6 & 1 & 0 & 0 & 0 & 0 & 0 & 0 \\ 0 & 0 & 1 & .6 & 0 & 0 & 0 & 0 \\ 0 & 0 & .6 & 1 & 0 & 0 & 0 & 0 \\ 0 & 0 & 0 & 0 & 1 & .6 & 0 & 0 \\ 0 & 0 & 0 & 0 & .6 & 1 & 0 & 0 \\ 0 & 0 & 0 & 0 & 0 & 0 & 1 & .6 \\ 0 & 0 & 0 & 0 & 0 & 0 & .6 & 1 \end{bmatrix} \quad (5.3)$$

The simulated data has panel errors **u** with actual correlation given by

$$\begin{bmatrix} 1.00 & 0.57 & 0.07 & 0.00 & 0.02 & 0.00 & 0.01 & 0.02 \\ 0.57 & 1.00 & 0.02 & 0.06 & -0.03 & 0.02 & 0.04 & 0.02 \\ 0.07 & 0.02 & 1.00 & 0.63 & -0.04 & -0.02 & 0.05 & 0.10 \\ 0.00 & 0.06 & 0.63 & 1.00 & 0.00 & 0.10 & -0.20 & 0.01 \\ 0.02 & -0.03 & -0.04 & 0.00 & 1.00 & 0.62 & -0.03 & 0.05 \\ 0.00 & 0.02 & -0.02 & 0.10 & 0.62 & 1.00 & -0.00 & 0.11 \\ 0.01 & 0.04 & 0.05 & -0.20 & -0.03 & -0.00 & 1.00 & 0.58 \\ 0.02 & 0.02 & 0.10 & 0.01 & 0.05 & 0.11 & 0.58 & 1.00 \end{bmatrix} \quad (5.4)$$

This is a balanced panel dataset with 10 panels of 8 observations each. A description of the panel structure of the data is:

```
         id:  1, 2, ..., 10                                    n =         10
          t:  1, 2, ..., 8                                     T =          8
              Delta(t) = 1; (8-1)+1 = 8
              (id*t uniquely identifies each observation)

Distribution of T_i:   min      5%     25%      50%     75%     95%      max
                         8       8       8        8       8       8        8

            Freq.  Percent    Cum. |  Pattern
         ---------------------------+----------
              10    100.00  100.00 |  11111111
         ---------------------------+----------
              10    100.00         |  XXXXXXXX
```

A summary of the variables is:

```
Variable         |      Mean   Std. Dev.       Min        Max |    Observations
-----------------+--------------------------------------------+----------------
y        overall |  2.184972   1.304209  -.9349678   4.802304 |     N =      80
         between |             .464511   1.347499   2.932273  |     n =      10
         within  |            1.226503  -.8024338   4.276666  |     T =       8
                 |                                            |
x1       overall |    .3625    .4837551          0          1 |     N =      80
         between |             .1094494       .125         .5 |     n =      10
         within  |             .472336      -.1375     1.2375 |     T =       8
                 |                                            |
x2       overall |    .7625    .428236           0          1 |     N =      80
```

DATASETS

between	|	.1608355	.5	1 |	n =	10
within	|	.3997626	-.1125	1.2625 |	T =	8

The data are:

id	t	y	x1	x2
1	1	-.9349678	0	0
1	2	3.023944	1	1
1	3	.7980191	0	0
1	4	1.952786	0	1
1	5	2.501896	1	1
1	6	1.749945	0	0
1	7	3.424954	0	1
1	8	3.902931	1	1
2	1	1.056602	1	0
2	2	1.068845	0	1
2	3	-.709373	0	0
2	4	-.1086365	0	0
2	5	3.485659	0	1
2	6	3.874307	1	1
2	7	3.175718	1	1
2	8	3.673608	1	1
3	1	.1077769	0	1
3	2	1.806073	1	1
3	3	.7593862	0	1
3	4	.6223395	0	0
3	5	2.293534	0	1
3	6	3.23475	1	1
3	7	1.085158	1	1
3	8	.8709748	0	0
4	1	2.686399	0	1
4	2	1.172582	0	0
4	3	2.183243	0	1
4	4	.706732	0	1
4	5	.6680957	0	0
4	6	2.941916	1	1
4	7	3.103627	1	1
4	8	1.280557	0	1
5	1	2.400463	0	1
5	2	2.498784	0	1
5	3	2.436319	0	1
5	4	1.758729	1	0
5	5	3.310097	0	1
5	6	4.549758	1	1
5	7	.9309282	1	0
5	8	2.746981	0	1
6	1	2.883902	0	1
6	2	3.710844	0	1
6	3	3.233667	1	1
6	4	3.014215	0	1
6	5	3.004494	1	1
6	6	2.191473	0	1
6	7	1.262497	0	1
6	8	4.157091	1	1
7	1	2.063806	1	1
7	2	.5553885	0	1
7	3	3.634825	1	1
7	4	3.375006	1	1
7	5	1.18627	0	1
7	6	3.41769	1	1

7	7	.068189	0	1
7	8	2.234942	0	1
8	1	2.310676	1	1
8	2	.7103744	0	1
8	3	3.370353	0	1
8	4	2.050668	0	1
8	5	2.348818	0	1
8	6	1.36899	0	0
8	7	4.722698	1	1
8	8	4.802304	1	0
9	1	.7837571	0	1
9	2	2.188972	1	1
9	3	4.319549	0	1
9	4	2.956809	0	0
9	5	3.090703	0	1
9	6	2.731381	0	1
9	7	1.131808	0	1
9	8	1.467309	0	1
10	1	3.704787	1	1
10	2	.1794717	0	0
10	3	1.384011	0	1
10	4	3.720814	1	1
10	5	3.17122	1	1
10	6	1.828142	0	0
10	7	.9966773	0	0
10	8	1.37176	0	0

5.2.6 Simulated measurement error data for the PA-GEE

In order to illustrate techniques for constructing the sandwich estimate of variance for a two-step estimator, we simulated data for a linear regression model given by

$$y_i = \beta_0 + \beta_1 x_{1i} + \beta_2 x_{2i} + \beta_3 x_{3i} \tag{5.5}$$

However, $\mathbf{x}_3$ is unobserved. Instead, we have $\mathbf{w}$ which is equal to the unobserved variable plus error and an instrumental variable $\mathbf{s}$.

A description of the panel structure of the data is:

```
        id:  1, 2, ..., 10                              n =     10
        t:   1, 2, ..., 4                               T =      4
             Delta(t) = 1; (4-1)+1 = 4
             (id*t uniquely identifies each observation)

Distribution of T_i:   min      5%     25%     50%     75%     95%     max
                         4       4       4       4       4       4       4

     Freq.  Percent    Cum. |  Pattern
     ---------------------------+---------
        10   100.00   100.00 |  1111
     ---------------------------+---------
        10   100.00          |  XXXX
```

A summary of the variables is:

DATASETS

Variable		Mean	Std. Dev.	Min	Max	Observations	
y	overall	21.97299	10.0514	4.094059	41.2298	N =	40
	between		5.700181	15.44533	31.22504	n =	10
	within		8.428398	4.086007	39.39337	T =	4
x1	overall	2.975	1.329883	1	5	N =	40
	between		.7115125	2	4.25	n =	10
	within		1.140738	.725	5.725	T =	4
x2	overall	5.4	3.248668	1	10	N =	40
	between		2.038518	2.75	7.75	n =	10
	within		2.591901	-.85	9.9	T =	4
w	overall	-.2024376	1.154352	-2.57972	2.093424	N =	40
	between		.4802081	-1.085854	.4280724	n =	10
	within		1.058143	-2.665788	1.706574	T =	4
s	overall	-.2481142	.9689245	-2.152968	1.875231	N =	40
	between		.505154	-1.220872	.2404079	n =	10
	within		.8386081	-2.64149	1.386709	T =	4

The data are:

id	y	x1	x2	w	s
1	31.744459	4	9	-1.2081616	-1.0708159
1	26.469884	4	4	1.4180836	1.0828664
1	33.265919	1	10	-.77037395	-.1847518
1	33.419892	4	8	.4144579	.23857762
2	29.936009	3	9	-1.1150387	-.75967723
2	37.912047	4	9	.57584045	.59421399
2	18.311498	2	3	1.3539961	1.1312023
2	27.690159	4	7	-.78054902	-.79222115
3	14.424474	2	2	.75987514	.54007814
3	30.411879	1	6	1.7559449	1.8752313
3	4.0940588	3	2	-2.0352778	-2.1529676
3	18.323714	5	1	1.2317474	.69928968
4	29.770692	3	9	-1.6566934	-.81543436
4	34.045546	4	9	-.81998911	-.43137562
4	41.229802	4	10	1.1209088	.45670796
4	11.166042	2	1	2.0934245	1.5466827
5	6.8228099	3	3	-1.6398477	-2.1046986
5	16.084249	2	4	.62104332	.03954247
5	31.451515	3	10	-1.1174939	-1.1654088
5	24.366558	3	7	-.51899332	-1.0290361
6	12.571442	1	5	-.88550458	-1.2184117
6	30.272789	1	10	-.53659636	-.44657953
6	18.958487	2	7	-1.9905849	-1.3242139
6	29.157948	4	9	-.93072945	-1.8942821
7	8.1158695	2	1	-1.1448774	-.01908745
7	26.118726	3	9	-2.5797199	-1.8719163
7	11.07891	2	2	.55967331	.15424293
7	38.331683	2	10	.71729918	.67129126
8	18.609577	5	3	-1.1565446	-.5663126
8	22.598339	2	6	.14885816	-.23900719
8	16.840708	5	2	-.50131764	.08000681
8	16.598596	5	3	-.15275318	-.59266891
9	9.4027453	4	2	-.71003654	-1.2672068
9	11.152214	2	2	.81480424	.22284953
9	13.69361	4	1	1.1922808	.54590719

9	32.695215	5	7	-.13302718	.05994656
10	12.419609	1	4	-.05218049	-.3663847
10	16.27943	1	5	-.82370427	-.28959267
10	25.63903	5	4	1.3528266	.82080319
10	7.4432654	2	1	-.96857179	-.08195498

References

Belsey, D. A., Kuh, E., and Welsch, R. E. 1980. *Regression Diagnostics: Identifying Influential Data and Sources of Collinearity.* New York: Wiley.

Bieler, G. and Williams, R. 1997. *Analyzing Repeated Measures and Cluster-Correlated Data using SUDAAN, Release 7.5.* Research Triangle Park, NC: Research Triangle Institute.

Billingsley, P. 1986. *Probability and Measure* (Second edition). New York: Wiley.

Binder, D. A. 1983. On the variances of asymptotically normal estimators from complex surveys. *International Statistical Review* 51: 279–292.

———. 1992. Fitting Cox's proportional hazards models from survey data. *Biometrika* 79(1): 139–147.

Bland, J. M. and Altman, D. G. 1986. Statistical methods for assessing agreement between two methods of clinical treatment. *Lancet* I: 307–310.

Carey, V. J., Zeger, S. L., and Diggle, P. J. 1993. Modelling multivariate binary data with alternating logistic regressions. *Biometrika* 80: 517–526.

Carroll, R. J. and Kauermann, G. to appear. The sandwich variance estimator: Efficiency properties and coverage probability of confidence intervals. *Journal of the American Statistical Association.*

Carroll, R. J. and Pederson, S. 1993. On robustness in the logistic regression model. *Journal of the Royal Statistical Society - Series B* 55: 693–706.

Chang, Y.-C. 2000. Residuals analysis of the generalized linear models for longitudinal data. *Statistics in Medicine* 19: 1277–1293.

Dempster, A. P., Laird, N. M., and Rubin, D. B. 1977. Maximum likelihood estimation from incomplete data via the EM algorithm (with discussion). *Journal of the Royal Statistical Society - Series B* 39: 1–38.

Diggle, P. J. and Kenward, M. G. 1994. Informative dropout in longitudinal data analysis (with discussion). *Applied Statistics* 43: 49–94.

Diggle, P. J., Liang, K.-Y., and Zeger, S. L. 1994. *Analysis of Longitudinal Data.* Oxford OX2 6DP: Oxford University Press.

Fanurik, D., Zeltzer, L. K., Roberts, M. C., and Blount, R. L. 1993. The relationship between children's coping styles and psychological interventions for cold pressor pain. *Pain* 53: 213–222.

Feiveson, A. H. 1999. What is the delta methods and how is it used to estimate the standard error of a transformed parameter? http://www.stata.com/support/faqs/stat/deltam.html.

Fitzmaurice, G. M., Laird, N. M., and Lipsitz, S. R. 1994. Analysing incomplete longitudinal binary responses: A likelihood-based approach. *Biometrics* 50: 601–612.

Gill, P. E., Murray, W., and Wright, M. H. 1981. *Practical Optimization.* New York: Academic Press.

Gould, W. and Sribney, W. 1999. *Maximum Likelihood Estimation with Stata.* College Station, TX: Stata Press.

Gourieroux, C. and Monfort, A. 1993. Pseudo-likelihood methods. In G. Maddala, C. Rao, and H. Vinod (eds.), *Handbook of Statistics*, Vol. 11.

Greene, W. 2000. *Econometric Analysis* (Fourth edition). Upper Saddle River, New Jersey: Prentice Hall.

Hall, D. B. 2001. On the application of extended quasilikelihood to the clustered data case. *The Canadian Journal of Statistics* 29(2): 1–22.

Hall, D. B. and Severini, T. A. 1998. Extended generalized estimating equations for clustered data. *Journal of the American Statistical Association* 93: 1365–1375.

Hardin, J. W. and Hilbe, J. M. 2001. *Generalized Linear Models and Extensions*. College Station: Stata Press.

Haslett, J. 1999. A simple derivation of deletion diagnostic results for the generalized linear model with correlated errors. *Journal of the Royal Statistical Society - Series B* 61(3): 603–609.

Heyting, A., Tolboom, J. T. B. M., and Essers, J. G. A. 1992. Statistical handling of drop-outs in longitudinal clinical trials. *Statistics in Medicine* 11: 2043–2062.

Hilbe, J. M. 1993a. Generalized linear models. *Stata Technical Bulletin* 11: 20–28.

———. 1993b. Log negative binomial regression as a generalized linear model. *Graduate College Committee on Statistics*.

———. 1994a. Generalized linear models. *The American Statistician* 48(3): 255–265.

———. 1994b. Log negative binomial regression using the GENMOD procedure in SAS/STAT software. *SUGI* pp. 1199–1204.

———. 1994c. Negative binomial regression. *Stata Technical Bulletin* 18: 2–5.

Horton, N. J., Bebchuk, J. D., Jones, C. L., Lipsitz, S. R., Catalano, P. J., Zahner, G. E. P., and Fitzmaurice, G. M. 1999. Goodness-of-fit for GEE: An example with mental health service utilization. *Statistics in Medicine* 18: 213–222.

Horton, N. J. and Lipsitz, S. R. 1999. Review of software to fit generalized estimating equation regression models. *The American Statistician* 53: 160–169.

Hosmer Jr., D. W. and Lemeshow, S. 1980. Goodness-of-fit tests for the multiple logistic regression model. *Communications in Statistics* A9: 1043–1069.

Huber, P. J. 1967. The behavior of maximum likelihood estimates under nonstandard conditions. In *Proceedings of the Fifth Berkeley Symposium on Mathematical Statistics and Probability*, Vol. 1, pp. 221–233, Berkeley, CA. University of California Press.

Karim, M. R. and Zeger, S. L. 1989. A SAS macro for longitudinal data analysis. *Department of Biostatistics, The Johns Hopkins University*: Technical Report 674.

Lee, A., Scott, A., and Soo, S. 1993. Comparing Liang–Zeger estimates with maximum likelihood in bivariate logistic regression. *Statistical Computation and Simulation* 44: 133–148.

Lesaffre, E. and Spiessens, B. 2001. On the effect of the number of quadrature points in a logistic random-effects model: An example. *Applied Statistics* 50: 325–335.

Liang, K.-Y. and Zeger, S. L. 1986. Longitudinal data analysis using generalized linear models. *Biometrika* 73: 13–22.

Lin, D. Y. and Wei, L. J. 1989. The robust inference for the Cox proportional hazards model. *Journal of the American Statistical Association* 84(408): 1074–1078.

Lin, L. I.-K. 1989. A concordance correlation coefficient to evaluate reproducibility. *Biometrics* 45: 255–268.

Lindsey, J. K. 1997. *Applying generalized linear models*. Berlin: Springer-Verlag.

Lipsitz, S. R., Fitzmaurice, G. M., Orav, E. J., and Laird, N. M. 1994. Performance

REFERENCES

of generalized estimating equations in practical situations. *Biometrics* 50: 270–278.

Lipsitz, S. R., Laird, N. M., and Harrington, D. P. 1992. A three-stage estimator for studies with repeated and possibly missing binary outcomes. *Applied Statistics* 41(1): 203–213.

Little, R. J. A. 1988. A test of missing completely at random for multivariate data with missing values. *Journal of the American Statistical Association* 83(404): 1198–1202.

———. 1995. Modeling the drop-out mechanism in repeated-measures studies. *Journal of the American Statistical Association* 90(431): 1112–1121.

Little, S. R. and Rubin, D. B. 1987. *Statistical Analysis with Missing Data*. New York: Wiley.

McCullagh, P. and Nelder, J. A. 1989. *Generalized Linear Models* (Second edition). London: Chapman & Hall.

McKusick, L., Coates, T. J., Morin, S. F., Pollack, L., and Hoff, C. 1990. Longitudinal predictors of reductions in protected anal intercourse among gay men in San Francisco: The AIDS Behavioral Research Project. *American Journal of Public Health* 80: 978–983.

Nelder, J. A. and Pregibon, D. 1987. An extended quasi-likelihood function. *Biometrika* 74: 221–232.

Nelder, J. A. and Wedderburn, R. W. M. 1972. Generalized linear models. *Journal of the Royal Statistical Society - Series A* 135(3): 370–384.

Neuhaus, J. M. 1992. Statistical methods for longitudinal and clustered designs with binary responses. *Statistical Methods in Medical Research* 1: 249–273.

Pan, W. 2001a. Akaike's information criterion in generalized estimating equations. *Biometrics* 57: 120–125.

———. 2001b. On the robust variance estimator in generalised estimating equations. *Biometrika* 88(3): 901–906.

Preisser, J. S. and Qaqish, B. F. 1996. Deletion diagnostics for generalized estimating equations. *Biometrika* 83: 551–562.

———. 1999. Robust regression for clustered data with application to binary responses. *Biometrics* 55: 574–579.

Prentice, R. L. and Zhao, L. P. 1991. Estimating equations for parameters in means and covariances of multivariate discrete and continuous responses. *Biometrics* 47: 825–839.

Rabe-Hesketh, S., Skrondal, A., and Pickles, A. 2002. Reliable estimation of generalized linear mixed models using adaptive quadrature. *The Stata Journal* 1(3).

Robins, J. M., Rotnitzky, A. G., and Zhao, L. P. 1995. Analysis of semiparametric regression models for repeated outcomes in the presence of missing data. *Journal of the American Statistical Association* 90(429): 106–121.

Rotnitzky, A. and Jewell, N. P. 1990. Hypothesis testing of regression parameters in semiparametric generalized linear models for cluster correlated data. *Biometrika* 77(3): 485–497.

Rotnitzky, A. and Robins, J. M. 1995. Semiparametric regression estimation in the presence of dependent censoring. *Biometrika* 82(4): 805–820.

Rotnitzky, A. G. and Wypij, D. 1994. A note on the bias of estimators with missing data. *Biometrics* 50: 1163–1170.

Rubin, D. B. 1976. Inferrence and missing data. *Biometrika* 63: 581–592.

SAS Institute, Inc.. 2000. *SAS/STAT User's Guide*. Cary, NC: SAS Institute Inc.

Shah, B. V., Barnwell, B. G., and Bieler, G. S. 1997. *SUDAAN User's Manual, Release 7.5*. Research Triangle Park, NC: Research Triangle Institute.

Shih, W. J. 1992. On informative and random dropouts in longitudinal studies. *Biometrics*: 970–971.

Sribney, W. M. 1999. What is the difference between random-effects and population-averaged estimators? http://www.stata.com/support/faqs/stat/repa.html.

Sutradhar, B. C. and Das, K. 1999. On the efficiency of regression estimators in generalised linear models for longitudinal data. *Biometrika* 86(2): 459–465.

Thall, P. F. and Vail, S. C. 1990. Some covariance models for longitudinal count data with overdispersion. *Biometrics* 46: 657–671.

Wacholder, S. 1986. Binomial regression in GLIM: Estimating risk ratios and risk differences. *American Journal of Epidemiology* 123(1): 174–184.

Ware, J. H., Docker III, S. A., Speizer, F. E., and Ferris Jr., B. G. 1984. Passive smoking, gas cooking, and respiratory health of children living in six cities. *American Review of Respiratory Diseases* 29: 366–374.

Wedderburn, R. W. M. 1974. Quasi-likelihood functions, generalized linear models, and the Gauss–Newton method. *Biometrika* 61(3): 439–447.

Zeger, S. L. and Karim, M. R. 1991. Generalized linear models with random effects; a Gibbs sampling approach. *Journal of the American Statistical Association* 86(413): 79–86.

Zhao, L. P. and Prentice, R. L. 1990. Correlated binary regression using a quadratic exponential model. *Biometrika* 77: 642–648.

Zheng, B. 2000. Summarizing the goodness of fit of generalized linear models for longitudinal data. *Statistics in Medicine* 19: 1265–1275.

Ziegler, A., Kastner, C., Grömping, U., and Blettner, M. 1996. The generalized estimating equations in the past ten years: An overview and a biomedical application. ftp://ftp.stat.uni-muenchen.de/pub/sfb386/paper24.ps.Z.

Author index

Altman, D.G., 166

Barnwell, B.G., 106, 130
Bebchuk, J.D., 167
Belsey, D.A., 158
Bieler, G.S., 14, 106, 130
Billingsely, P., 29
Binder, D.A., 33
Bland, J.M., 166
Blettner, M., 132
Blount, R.L., 174

Carey, V.J., 85, 89
Carroll, R.J., 51, 120
Catalano, P.J., 167
Chang, Y.-C., 143
Coates, T.J., 134

Das, K., 86
Dempster, A.P., 128
Diggle, P.J., 89, 128, 162
Docker III, S.A., 192

Essers, J.G.A., 128

Fanurik, D., 174
Feiveson, A.H., 70
Ferris Jr., B.G., 192
Fitzmaurice, G.M., 89, 128, 167

Gill, P.E., 18
Gould, W.W., 7
Gourieroux, C., 105
Grömping, U., 94, 132

Hall, D.B., 117, 119, 131
Hardin, J.W., 5, 6, 8, 11, 18, 27, 51, 65, 69, 87, 130
Harrington, D.P., 95
Haslett, J., 159
Heyting, A., 128

Hilbe, J.M., 5, 6, 8, 11–13, 18, 24, 27, 28, 51, 65, 69, 87, 130
Hoff, C., 134
Horton, N.J., 12, 167
Hosmer Jr., D.W., 167
Huber, P., 28

Jewell, N.P., 170
Jones, C.L., 167

Karim, M.R., 49, 85
Kastner, C., 132
Kauermann, G., 51
Kenward, M.G., 128
Kuh, E., 158

Laird, N.M., 89, 95, 128
Lee, A., 31
Lemeshow, S.L., 167
Liang, K.-Y., 3, 5, 57, 94, 119, 130–132, 162
Lin, D.Y., 31
Lin, L.I.-K., 166
Lindsey, J.K., 28
Lipsitz, S.R., 12, 89, 95, 128, 167
Little, R.J.A., 127, 128, 176

McCullagh, P., 3, 24, 28
McKusick, L., 134
Monfort, A., 105
Morin, S.F., 134
Murray, W., 18

Nelder, J.A., 3, 28, 117
Neuhaus, J.M., 49, 134

Orav, E.J., 89

Pan, W., 87, 134, 139
Pederson, S., 120
Pickles, A., 47
Pollack, L., 134

Pregibon, D., 117
Preisser, J.S., 119, 159
Prentice, R.L., 105

Qaqish, B.F., 119, 159

Rabe-Hesketh, S., 47
Roberts, M.C., 174
Robins, J.M., 127
Rotnitzky, A.G., 123, 124, 127, 170
Rubin, D.B., 123, 128

Scott, A., 31
Severini, T.A., 117
Shah, B.V., 106, 130
Shih, W.J., 127
Skrondal, A., 47
Soo, S., 31
Speizer, F.E., 192
Sribney, W.M., 7, 56
Sutradhar, B.C., 86

Tolboom, J.T.B.M., 128

Wacholder, S., 26
Ware, J.H., 192
Wedderburn, R.W.M., 8, 24, 26, 55, 154
Wei, L.J., 31
Welsch, R.E., 158
Williams, R., 14
Wright, M.E., 18
Wypij, D., 123, 124

Zahner, G.E.P., 167
Zeger, S.L., 3, 5, 49, 57, 85, 89, 94, 119, 131, 132, 162
Zeltzer, L.K., 174
Zhao, L.P., 105, 127
Zheng, B., 165
Ziegler, A., 132

Subject index

AIC, 110, 139
ALR, 91
apparent overdispersion, 4

balanced panels, 63
BIC, 110
boxplot, 143

canonical link, 7
clustered data, 31
common correlation, 59
complete panel, 122
compound symmetry, 59
concordance correlation, 166
Cook's distance, 158, 165
cumulative logistic regression, 108

data
    clustered, 31
    independent, 17
    longitudinal, 32
    panel, 31
DFBETA, 158
DFFIT, 158
dispersion, 26
dropout, 123

EDA, 143
empirical variance, 29
equal correlation, 59
exchangeable correlation, 59

Fisher scoring, 6, 138
fixed effects
    conditional, 36
    unconditional, 34

gamma model, 65
geex, 12
geometric model, 69
GLS, 63
goodness of fit, 167

Hessian, 28

imputation, 123
independence model, 57
independent data, 17
IRLS, 138
item nonresponse, 123

jitter, 147

leverage, 165
likelihood ratio test, 169
link function, 26
longitudinal data, 32

MAR, 123
marginal model, 49
MCAR, 123
missing data
    complete panel, 122
    dropout, 123
    informatively, 123
    item nonresponse, 123
    panel nonresponse, 122
    patterns, 122
    process, 123
model
    marginal, 49
    population-averaged, 49
    subject-specific, 49
multinomial logit, 106

naive test, 170
Newton–Raphson, 6

odds, 90
odds ratio, 90

panel data, 31
panel level covariate, 130
panel nonresponse, 122
patterns

missing data, 122
population average, 56
population-averaged model, 49

QIC, 139
QIC_u, 142
quadrature
    adaptive, 47
    Gauss–Hermite, 45
quasilikelihood, 27

random effects, 42
relative risk ratio, 106
residuals
    scores, 29

sandwich, 28
score test, 169
scores, 29
singleton, 88
subject specific, 56
subject-specific model, 49

test
    generalized, 169
    likelihood ratio, 169
    naive, 170
    score, 169
    Wald, 169
    Wald–Wolfowitz, 143
    working, 170

variance
    Hessian, 28
    modified sandwich, 30
    sandwich, 28

Wald test, 169
Wald–Wolfowitz test, 143

YAGS, 12, 85